The Herasaga

Book One: Hera, or Empathy

Book Two: The Priesthood of Science

Book Three: Hera the Buddha

Reader's Comments on Hera, or Empathy

Hera, or Empathy is a profound work of literature. Leiss does a remarkable job of pulling together various issues that face humanity today, weaving a slightly futuristic tale that entices the reader to ponder the actions of mankind—past, present, and future—and to consider the disaster that could be waiting for us on the other side of science.

Josh P. McClary, Chicago, Illinois
Author of *Chance Murphy and the Battle of Morganville*

This novel is a compelling read for anyone who cares to think about the implications to the nature of being human that arise from our unrestrained manipulation of nature driven by the accelerating advances of scientific knowledge.

Steve E. Hrudey, FRSC
University of Alberta

I loved it, especially the compelling debates among the characters. The clash between religion and science is such a timely topic.

Lee Merkhofer
Cupertino, California

In *Hera, or Empathy,* Bill Leiss has written an engrossing and compelling novel of ideas. In doing so, he has confronted religion with science, and the basic tenets of both are the subject of beautifully written logical dialogues causing the reader to reflect deeply upon what it means to be human and what should be the limits of science.

Prof. Benjamin Smith (ret.)
École Polytechnique, Montréal

Hera is a major contribution to the debate concerning the relation between humanity and nature which is bound to become a reference point for both the near and far future.

Professor Ian Angus
Simon Fraser University

THE Priesthood of Science

A WORK OF UTOPIAN FICTION

WILLIAM LEISS

A Cangrande Book

A Cangrande Book is a University of Ottawa Press imprint.

A Cangrande Book™
Ottawa, Canada
www.uopress.uottawa.ca

This is a work of fiction. Names, characters, places, and incidents are the products of the author's imagination or are used fictitiously. Any resemblance to actual persons, living or dead, is entirely coincidental.

The University of Ottawa Press acknowledges with gratitude the support extended to its publishing list by the University of Ottawa, by Heritage Canada through its Book Publishing Industry Development Program, and by the Canada Council for the Arts.

Library and Archives Canada Cataloguing in Publication

Leiss, William, 1939-
The Priesthood of Science : a work of Utopian fiction / William Leiss.

(The Herasaga ; bk. 2)
"A Cangrande Book".
ISBN 978-0-7766-0677-4

I. Title. II. Series: Leiss, William, 1939- Herasaga ; bk. 2.

PS8623.E475P75 2008 C813'.6 C2008-902083-9

Cover artwork: Alex Colville (Canadian b. 1920), *Horse and Train* (1954), glazed tempera on masonite. Art Gallery of Hamilton, Gift of Dominion Foundries and Steel, Ltd. (DOFASCO Inc.), 1957

Edited by Jenny Gates
Cover and text designed by Hydesmith Communications

PRINTED & BOUND IN CANADA

Dedicated to the memory of my mother

This much is certain:

If, improbable as it seems, the universe we inhabit was created by a just and righteous Supreme Being,

And if, yet more improbably still, that Being harvests after death the souls of those on this earth who had led blameless and self-sacrificing lives,

Depositing them in a place of sweet repose that goes by the name of Heaven, there to be serenaded by angels for all eternity,

then my mother is in Heaven.

The end of our foundation is the knowledge of causes, and secret motions of things; and the enlarging of the bounds of human empire, to the effecting of all things possible.

And this we do also: we have consultations, which of the inventions and experiences which we have discovered shall be published, and which not: and take all an oath of secrecy for the concealing of those which we think fit to keep secret: though some of those we do reveal sometime to the state, and some not.

Francis Bacon, *New Atlantis* (1627)

Even when one selects a method of making a living which is independent of the search for knowledge, one must then also decide to keep one's knowledge to oneself, or to interchange ideas only privately amongst friends, as was customary during the 17TH and 18TH centuries, for otherwise others are still going to misuse the results for evil purposes, and I feel that one would then never be free of responsibility.

Max Born, Letter to Albert Einstein (1955)

All of our exalted technological progress, civilization for that matter, is comparable to an axe in the hand of a pathological criminal.

Albert Einstein, Letter to Heinrich Zangger (1917)

I read in the paper recently that you are supposed to have said: "If I were to be born a second time, I would become not a physicist, but an artisan." These words were a great comfort to me, for similar thoughts are going through my mind as well, in view of the evil which our once so beautiful science has brought upon the world.

Max Born, Letter to Albert Einstein (1954)

The Priesthood of Science

A WORK OF UTOPIAN FICTION

Contents

PROLOGUE:

"BEGOTTEN, NOT MADE"

Reading from her notebooks, my stepmother Hera had recounted this dream of hers to me some years ago, to considerable hilarity on her part, as well as on mine, shortly after my return to earth from my aborted mission to Mars. I stumbled on it again whilst reviewing the assortment of documents I had assembled in order to compile this volume, along with the recording I made of our subsequent discussion:

> The silent crowd of pilgrims sitting cross-legged on the desert floor in the shadow of Bare Mountain numbered about thirty thousand. They had made their way there on foot, except for the young children and the disabled who had ridden with the group's supplies in mule-drawn covered wagons. All had been refreshed with water and a simple meal shortly after having arrived at their destination at the end of their two-hundred-mile journey.
>
> Their route northward and westward had taken them from the outskirts of Bakersfield, California, where the Sujana Foundation Hospice is located, first to the grounds of the Lake Isabella Pavilion, which the Foundation operates as a nursing home and palliative care facility. There they had rested briefly and built up their strength before striking out on the most arduous segment of the pilgrimage, first turning south and following Route 178 over the 5,000-feet Walker Pass through the Sierra Nevada mountain range, then swinging north again through Indian Wells Valley by China Lake, through the Argus Range and on into Panamint

Valley. The long column was then in the desert and moving more slowly as it entered Death Valley, heading north, still using Route 178 to Emigrant Junction, where it swung west again on Route 190.

The deserts in the southwestern United States bloom in springtime, a three-month profusion of riotous color among the grasses, cacti, shrubs, and annuals, including those that compete for the attention of pollinating insects in the brief interlude before the brutal summer heat strikes. Amongst the shrubs there is the creosote bush, with flowers of brilliant yellow; the indigo bush, violet or deep purple; the desert holly, flowering greenish yellow, with leaves that later turn a silvery gray; and the Death Valley sage, combining purplish flowers and white seedcases. There are barrel cacti and the calico cactus, as well as the beavertail cactus with its extraordinary magenta flowers in springtime.

The flowering plants, which hunker down at ground level during dry years and triple in size in the wetter ones, include many varieties of evening primrose, the gold poppy, desert heliotrope, whispering bells, and two kinds of daisy—desert ghost and desert star. At higher elevations on the surrounding hills and mountain ranges are the trees—pinions, juniper, mountain mahogany, and bristlecone pine—while around springs and watercourses on the desert floor grow cottonwoods, willows, velvet mesquite, and screwbean mesquite trees. The pilgrims were in awe as they passed by these hardy survivors ensconced in the forbidding landscape.

When at last the column reached Stovepipe Wells, they were at sea level, or what used to be sea level before the seas had started rising a half century ago. They transited the Funeral Mountains and entered the Amargosa Desert, finally reaching the spot where the little town of Beatty, Nevada, once stood, and where posted signs told them that they were nearing their destination. A few miles further on, around midday, the dry and dusty contingent stumbled into the encampment prepared for them on the western edge of Crater Flat, a valley floor sandwiched between Bare Mountain to the west and Yucca Mountain to the east.

Onward from Lake Isabella, their column had been guided at the front by a group of ministers from the Church of the Red Planet, identified by the simple logo they wore—a reddish-pink orb. Then, at the spot along Route 178 where the pilgrims reached the entrance to Death Valley National Monument, the column had been met by officials from the Yucca Settlement, who were identified by their own distinctive insignia—a circular design showing a great mountain peak against a background of yellow triangles, the symbol of radioactivity hazard. The officials were there to offer the pilgrims entry, for the Settlement's boundary now extended westward to the foothills of the Sierra Nevada range, and passage through its territory was prohibited unless a transit permit had been granted.

The pilgrims then rested their weary limbs. Toward the end of the afternoon, their preachers had conducted the first of the series of ceremonies marking the Festival of the Prophet, which would culminate three days hence with a sumptuous vegetarian banquet. Throughout the afternoon and on into the early evening, however, the crowd's anticipation mounted steadily. Their gaze was fixed on

> the crest of Bare Mountain, where the spires of the Mother Church stood. For they had been promised that on this night, the Prophet Marco would descend from Bare Mountain and walk among them.

I hardly knew what to make of this script since I'm no more religiously inclined, in the conventional sense, than she herself is. How might the hidden meaning of her dream be interpreted? The maxim known as Occam's Razor counsels us not to "increase, beyond what is necessary, the number of entities required to explain anything." The multitude of pilgrims in her little parable might be just the type of superfluous entities that the estimable philosopher warns us to be on guard against. So is there a simpler explanation on offer? Maybe she just missed me during my brief journey away from planet earth.

No, there is a bit more to the business than that. Hera may not be a religious person, at least not in the eyes of true believers, but I do believe she is, and has been for some time, a part-time theologian. So when I asked her then what the vision meant to her, she replied, "I often have such dreams, Marco, because my mind is obsessed with Biblical imagery. You will recall, I'm sure, the protracted intellectual warfare I engaged in with my father when I challenged his facile assumption that he had the right to engineer his children. The creation story in the Book of Genesis was indispensable to me in helping me to form my own ideas for those battles.

"You may also recall how I used that story against him: That in his role as scientist and genetic engineer, he was operating as if he were a stand-in for the God of Genesis, who designs and implements His prelapsarian creation in a state of moral innocence. I told him that there was no such innocence anymore and that, metaphorically speaking, his acts, as well as those of his peers who were engaged in similar enterprises, could be regarded as the second stage of Original Sin. Or words to that effect."

"You will recall that I was present for many of those sessions," I replied, "and I remember well his profound puzzlement at what he once termed your 'bizarre religiosity.' But I don't think he ever really grasped why you found the religious ideas and imagery so compelling."

"Indeed you were; I needed you there for moral support. But I didn't reveal everything that was running through my mind at the time. When I colluded with him and my sisters in the creation of our Second Generation, these wonderful children now growing up in our midst, I did so with full awareness that as a new species they would carry this blemish—of having been engineered according to someone's arbitrary will—in their genes to the very end of their time on earth. To be sure, it isn't the same fault that other humans claim to bear within them—the one that was introduced when they defied their God's express command to forego the fruit of the Tree of the Knowledge of Good and Evil. And so, because the stain is of a different kind, our path to our own salvation must differ, too. What we Yuccans must do to expiate our sin struck me suddenly one day, when for some reason my mind had cycled back to some marvelous language in the Nicene Creed, one of the great texts of Christian faith: The passage says of Jesus that He was 'begotten, not made, being of one substance with the Father.'

"According to the Christian faith, all of creation was made, fashioned by God's hands—but Jesus was 'begotten.' Then I said to myself, yes, precisely, this is my faith: My kind, too, is begotten, not made, not worked to a plan by some disembodied deity. We are begotten of the earth, begotten of nature's evolutionary, randomized trials. We were not made for a purpose; we were not worked to the specifications of a plan. We emerged, spontaneously, accidentally, prepared for a destiny no different from that of all other nature-begotten species: To have our moment of glory and then to disappear forever, and in so doing to prepare the way for those new species who will follow us. I believe that it is our solemn duty to acquiesce in this shared fate."

"If I might interrupt your sermon for a moment," I said, laconically. "I am most anxious to hear about your path to salvation."

"I sense a whiff of sarcasm in your question, my dear son, but I will gratify your urge for enlightenment nonetheless, since you so obviously require it. We were remade, slightly, to our own father's design, and this act cannot be undone; we cannot wash away the sin, but only expiate it. We will do so by caring for the earth, by helping to reverse humanity's cancerous dominion over all else. We will restore the balance in nature's creation, the once-generous spaces where others flourished free of human interference, so that they might do so again. Our fate is bound together with that of all living things because we carry in us genes whose origins predate the branching of plants and animals a billion years ago.

"But we will take special care of our sisters and brothers, the great ape species, because for our own salvation we must. We were begotten simultaneously with them; their natural makeup is virtually identical with ours. More to the point, we have a stake in their future, as do they in ours. We once branched from the chimpanzees five million years ago or so. What novel and wonderful relatives of ours might appear from further spontaneous branchings in the Hominina subtribe, where humans, gorillas, bonobos, and chimps co-exist? We have driven them right to the edge of extinction because we covet every inch of the earth for our own needs. We have a duty to ensure that their kingdoms are restored to them. If we discharge this duty, as well as our larger responsibilities for the welfare of living nature, we may ask for forgiveness."

§ § §

The events recorded in the present volume open in March of 2064, seven years after Hera's whole tribe—eleven of the original sisters plus the cohort of more than a thousand youngsters, then aged eleven (what an interesting coincidence of numbers!)—had moved to their permanent quarters in and around Yucca Mountain in southern Nevada. By this time, the surrounding human settlements, including Las Vegas, that nearby self-caricature of an urban entity, had largely emptied out, victims of too little water and too much despair. Hera believed that the auguries were favorable for her tribe because, as she put it, they were following the wise maxim of the philosopher René Descartes, who lived during the seventeenth-century horrors precipitated by Europe's wars of religion: "He lives well who stays well hidden."

By the time my narrative's terminus is reached in the concluding chapter of this volume, barely five years will have elapsed, a short time in the grand scheme of things, but also one of great peril for the Yuccans, as Hera likes to call her people.

This is Book Two in the series I have dubbed *The Herasaga.* Here I have the honor of personally conducting the reader through every episode, a role to which I could not have aspired in the case of Book One, since I did not tumble into the world from my mother's womb until well after the action therein had commenced. For all those who cannot lay their hands upon the prior work, however, or alternatively, those who, having been seduced into forking over valuable coin for one product only to discover once the good is in their possession that extracting its value requires another purchase to be made, might then react to this imposition by plotting some suitable act of revenge upon the author; for those readers (should there be any), I offer here a brief précis of the volume I entitled *Hera, or Empathy,* a summary that may be bypassed, of course, by those lucky few who have treated themselves to the full meal.

The band of sisters, twelve in all, are the natural offspring of two parents, Ina Sujana and Franklin Stone, whose children had been conceived by means of *in vitro* fertilization using previously frozen eggs and sperm and subsequently brought to term in the wombs of surrogate Indonesian mothers. The sisters' parents, both famous neuroscientists, had planned in advance to carry out germline genetic modifications on the embryos—alterations in certain genes that would be inherited by any future offspring of their daughters. The modifications targeted genes responsible for the development of specific regions in the brain's prefrontal cortex, in particular the mysterious complex of neurological functions referred to as our sense of empathy. What they were hoping to achieve by "upregulating" these genes was to create a type of human being who would be strongly motivated to promote the common welfare of humankind.

All the sisters were born within about ten days of each other in the month of September 2014; Hera, the first-born of the lot, assumed a leadership role in the band from an early age. She was fifteen when she first learned what her parents had done to her and her sisters, and what her father's larger plans were. Their mother had died some years before they were born, and at the time they were separated from their father. After father and daughters were reunited, Hera began questioning him about his motivation in engineering his daughters, a topic charged with much emotion in itself, but made more so by her knowledge that he was determined to repeat the procedure on a far larger collection of embryos.

Her increasingly fractious colloquies with her father always left her unsatisfied. She held firmly to the view that anyone who, coming to the age of psychological maturity, stumbled upon the fact that she had been genetically designed to serve another's purpose, would be revolted by the very idea—although she could never persuade her father to concede the point. She battered him with arguments drawn from the Book of Genesis, the most widely known creation story of all, but, in her eyes, every session ended in failure because she simply couldn't dent the armor of his self-confidence in the triumph of modern science: "We know now how nature works; we are the new masters of creation; we have a duty to advance our understanding continuously and to turn our knowledge to the betterment of humankind, a task that includes the re-engineering of genomes, human, plant, and other animal, to relieve suffering and enhance our capacities. Yes, inevitably we will make mistakes, but we will learn from them as well, and we will utilize the powers bestowed on us by science to mitigate and remedy those mistakes."

She tried (and failed utterly) to make him see that rewiring the human brain was a radically different kind of enterprise from all other types of genetic manipulation. As she once explained to me, the still-unsolved mystery of this mighty organ is how a set of neurophysiological processes, operating by means of matter and energy—chemical and electrical signal transductions in nerve tissues, as captured by brain imaging devices—give rise to an immaterial "subjectivity," otherwise called our sense of self, which exerts an overwhelming emotional force upon us.

As she grew and matured, she knew this because she experienced in her own mind the consequences of the experiment he had performed on her brain. She believed she and her sisters did indeed have a much different sense of empathy as a result of what he had done, but it was not at all the same package of traits that had existed in her father's mind as he worked out the scientific protocol he then applied to the embryos. This ineffable subjectivity that blossoms spontaneously in the human brain, becoming in its full manifestation the glorious light of reason, she told him, is the most precious cargo we could ever receive from nature's evolutionary hand. We are not nearly wise enough to presume to tinker with it, she said then, and we may never be.

In attacking him, Hera was, however, caught firmly on the horns of an insoluble dilemma of her own making. She could not acquiesce in her father's future plans. She had resolved that, if he charged ahead with his scheme to engineer a new batch of embryos, she would find a way to sabotage his experiment. At the same time, she knew from introspection that she and her sisters embodied ways of feeling and thinking that set them apart from "normal" humans. Moreover—and here was the nub of the matter—she was deeply satisfied with what she was or had become, and she also knew that if, conceivably, she could be reverse-engineered, she would never agree to undergo the procedure! Thus the conundrum she faced was embedded in the simple fact of her own existence, and she could not resolve it. Left unanchored and adrift in her personal history, she felt herself sinking under the combined weight of her father's anger, her cycles of migraine headaches, and her bouts of mild depression.

She fought back against these forces and managed at least to hold them at bay. Then one day there came from the biologists among the sisters the revelation that they might be hybrids—a new variant of the species *Homo sapiens* that could not breed successfully with other humans. Eureka! A new species? The full implications of this awareness hit her like a thunderbolt. Somehow her father had done this to them, unbeknownst to him, and unintentionally, of course. And he was preparing to do it again, on a thousand embryos created from eggs and sperm donated by talented couples representing diverse genetic subtypes from every corner of the globe. Well, why not? Let's make both sexes this time on the assumption that they'll be fertile among themselves, and then we'll branch off from other humans and travel down our own path of evolutionary development. And so she rolled the dice and the great gamble commenced.

But along what path? Toward what goal? Would Hera's new tribe suffer a fate similar to that of the ancient Israelites, condemned by their God to wander aimlessly for forty years in the harsh Egyptian desert, longing for the promised land of milk and honey, all the while bitching at Moses about his leadership failures? At first it seemed so, at least to me. The forbidding Mojave Desert appeared to have a lot in common with the Biblical wilderness.

Hera's quest for the right path—in effect, her attempt to grasp in what way her life and mission were defined by her unique situation—began in the battles with her father. It's not surprising that in order to do so she turned for support to Mary Shelley's *Frankenstein,* notably to that great book's depiction of the conversations between Dr. Frankenstein and the mutilated entity he had brought to life. That nameless and tragic being tried and failed to make his creator understand the nature of the shared responsibility that bound the two of them together for all eternity. That being, whom Frankenstein had been wont to call "the demon," was in fact a new species, the first of its kind. In the creature's demand that the scientist should acknowledge the duty to extend his own mission to its logical end, by fashioning for him a female of his own kind, Frankenstein realized to his horror that the venture on which he had embarked now threatened to become another Adam-and-Eve story.

True, this had not been his original intention. Do you see how the situation of Hera's father mimics Dr. Frankenstein's? What the creature's demand had done was to expose the unpleasant truth that this bold scientist had simply failed to think through the full implications of his experiment with life. Once launched into the world, the demon initially had assumed that, despite his disfigurement, he was otherwise just an ordinary human being. He empathized with them and expected that the same feeling would be reciprocated. But those he encountered soon corrected this misconception; he would not be allowed by them to regard himself as a fellow creature.

Thus he drew the necessary conclusion from these bitter experiences and proposed that he should abandon Europe and live with his new mate entirely separate from humans, isolated "in the deserts of the new world," that is, in the Americas. Frankenstein inferred from this proposal that his demon wished to reproduce "true to type" and thus to propel his own unique genome into life's evolutionary gamble: Sink or swim, survive or perish. This story does not have a happy ending, as I am sure my readers are aware.

Hera and her sisters were no mutilated beings, to be sure, but starting fairly early on in life, they did begin to encounter internal feelings that led them to believe they were different somehow from other people, and that these differences arose accidentally or incidentally as a result of the gene manipulations their father had carried out on their brain functions. (Who can say what had actually happened in the development of their neurological tissue? Would any notable difference have shown up under magnetic resonance imaging, or under autopsy following death? Maybe not. What matters here is their own interpretation of the feelings that arose within them as they reflected on their life situation, including the awareness that their brains had been engineered.) If they were not "ordinary humans," then who or what were they?

Here again, subjectivity intervenes. Hera's reflections led her down a particular path and ended thus: We are products of modern science, or more fully, products of science's penetration of the processes of biological life and evolution, and its subsequent mastery over those processes. When she got there she said to herself, in effect: 'This is how and why I was created, namely, as an integral moment in science's will to grasp the secrets of organic life and to use that knowledge in new technologies. This act of human reason and will defines the uniqueness of my own being as a conscious entity on earth. I must therefore "appropriate," or fully absorb and internalize, my awareness of this act, considered as an intentional exercise of uniquely human powers, *as well as the feelings that this process of absorption induces within me.*'

She explained all this to me, very patiently I thought at the time, in a long colloquy that I recorded and reproduced in one of the later chapters in the previous volume. Much later it occurred to me that where she ended up was not preordained for her, nor had she arrived at the only possible terminus for such a voyage; in another's mind, or under other circumstances in life history, one might have docked elsewhere. But that was one of her main points, of course: For a reasoning animal, that is, an animal with a ratiocinative mind as well as a brain, such diverse possibilities are inherent in the situation.

Having placed her coming-into-being in the context that made the most sense to her, Hera concluded that conversation as follows: "I am terribly troubled by the process of my creation. What troubles me above all is the disjunction I see between the careful rationality of the scientific mode of reasoning, on the one hand, and the arbitrariness of the will that carried out this specific act of genesis, on the other. Under another type of protocol made feasible by science's mastery over biological processes, I might have been brought to life as a malevolent entity, designed to be an instrument in some despicable plan to sow horror and misery among humans or to subject them, on my creator's behalf, to bitter and eternal oppression. Using the words in which Frankenstein expresses his fears about what the demon and his mate might represent, once their kind had begun to reproduce, I myself could have been fashioned into a being 'who might make the very existence of the species of man a condition precarious and full of terror.' This thought is intolerable to me."

Such speculations supplied the motivation that drove Hera to read and reread that other—and competing—creation story, the one in the Book of Genesis, for the God of Genesis acts with what appears to be a similar arbitrariness of will during the six days of his labors. Theologians may invent all kinds of cute after-the-fact rationalizations for what supposedly happened then, but really, who can fathom the mind and will of God? *Im Anfang war die Tat*: "In the beginning was the deed," Goethe's Faust proclaims. Period. All the rest is mystery, because none of us is in a position to ask God to explain why She did the deed in the first place. Until modern science comes along to say, "There are no more mysteries about life on earth. We'll show you exactly how the deed was done—well, not completely, we don't yet have the full story, but we will, sooner or later; you can bank on it."

For the Christian believer, the story of the Apocalypse, as played out in the Book of Revelation, tells of the coming end of days. But in truth, the work of modern science is the real Apocalypse, in the literal meaning of that Greek term—the "lifting of the veil." Prior to science's revelations, the stage of life was concealed behind a double set of thick curtains; drawn upon the outer set, on the side facing the audience, was an elaborate and colorful painting entitled "the origins of life."

Depending on which version of the tapestry one worshipped before, there was a solo creator-god or many, male or female or hermaphrodite, supplemented by subsidiary supernatural beings of every conceivable type, benevolent or otherwise, represented as the authors of the work, and, filling the canvas, a bountiful outpouring of created entities. In the most famous representation of all, rendered in *buon fresco* style on the ceiling of the Vatican's Sistine Chapel, Michelangelo's God extends his arm toward Adam's outstretched limb in order to convey the spark of life across the gap separating their fingertips.

Now, the first set of curtains rises, and there on the stage stands fully arrayed for all to see the evolution of species and the descent of humankind. A many-limbed tree of life dominates the scene. Its mighty trunk floats in a hydroponic biochemical bath, the primeval soup, uniting every type of living thing—bacteria, viruses, plants, animals—in a common enterprise. Its branches reveal the secret of common ancestry among them as well as the chronology of their separations from each other.

In the audience, of course, all immediately search out *Homo sapiens sapiens*, their eyes moving rapidly past the point on the tree's limb where primates appear, then following the monkeys to the apes, finally past the orangutans and the gorillas, to the last branching on the hominoid limb, having arrived at the common ancestor of chimpanzees, bonobos, and humans. So compelling is this story to its adherents in the scientific community that this common ancestor can be presumed to have existed, some five million years ago, even though no fossil evidence confirms the hypothesis, because no alternative presumption could possibly explain the astonishing similarities in the genomes of those three species.

The mere mention of genomes causes the second, inner set of curtains to slowly rise, and as it does so, we slip beneath the surface of perceptual reality—the world as it appears to our senses of sight, smell, touch, and taste. Passing through the lenses of the scanning electron microscope, we enter the molecular kingdoms where proteins play. Shadowy fragments of DNA's breathtakingly beautiful double-helix ladder, employing just two steps, A/T and G/C, in infinite recombination, infiltrate the displays of genomic barcodes and chromosome readouts.

If what are called structural genes are regarded as the instruments in a musical ensemble, without which no sounds are possible, then the number and variety of instruments tell us what kinds of sounds any ensemble is capable of. And if the regulatory genes represent the players, who exercise control over when and how any particular instrument will be put into action, then it is the common score they all play from, written out over eons by each organism's evolved developmental blueprint, that instructs each of the musicians as to which notes to play, in what pitch, with what intervals, at what time, all in order to create a harmonious effect, a fully formed living creature in all its glorious singularity. Knowing this, how could one be surprised or puzzled upon hearing that the creatures which are so like us in molecular genetics—the chimpanzees and bonobos—could also be so unlike us in important ways?

In any event, the veils were lifted and the old storyline receded. But the expected applause was muted, and from the audience, shouts and murmurs welled up: "What do you think you're doing? Drop the curtains again! We like the old story a lot better than the new one! Besides, I'm no ape! And there'll be trouble if you try to tell me otherwise." After a while, in a good many lands, biologists were told in no uncertain terms to stay inside the locked doors of their laboratories, if they knew what was good for them, and to slip the new medical therapies they were paid to discover under their doors into the hallway, where the runners for pharmaceutical firms could retrieve them on their daily rounds and proceed to turn them into useful products. Under no circumstances were these scientists to be seen or heard in the neighborhood of any school where children were being instructed.

Hera became convinced that, when push came to shove, if most people should ever arrive at the point where they thought they must choose between religion and modern science, the result would be clear: Religion would win hands down. This prospect frightened her because she guessed there was a fair chance that some developments, especially in the manipulations of animal genomes, would eventually scare the wits out of the faithful, and, in their estimation, place their immortal souls in peril; as a result, they might decide they could do without any more medical miracles and proceed to shut down the scientific enterprise permanently. This possibility appalled her because she had earlier come to the conclusion that, as far as her new species was concerned, science was not just the progenitor of useful technologies. To them it would always and forever also have an existential meaning because it had shaped who they were meant to be as well as how they had come to be. As creatures designed by the scientific enterprise itself, they stood on its shoulders and saw the surrounding world of nature through its eyes and with its mental constructs. She has told the new generation of her kind that she hoped they would earn the right to brand themselves with a new scientific moniker: *Homo sapiens scientificus.*

At that point she had formulated not only her path forward and her ultimate goal, but also her grand plan as to how to get there. She went looking for an isolated and easily defendable permanent site for her tribe, and found it in the middle of the Mojave Desert, at Yucca Mountain. She obtained a complete library of world science on videodisc and salted it away, along with as much up-to-date scientific instrumentation as she could scavenge, inside secure tunnels next to the mountain's cache of nuclear waste. And, some years before the date when this present volume commences, fearing that the accelerating social chaos around the globe might, among other things, bring down in ruins, one by one, the world's premier scientific research institutes, she had asked her biologist sister Gaia to begin organizing a vast rescue mission in order to gather up leading scientists and their families, wherever they could be found, and install them in safe quarters at the private university they had founded in the then-vacant city of Las Vegas.

At last—and, I'm quite sure, not a moment too soon, in my reader's opinion—I am ready to resume my family saga once more. As the story opens, the finishing touches are being put on the core facilities at Yucca Settlement. It is self-sufficient in food, water, and necessities; increasingly isolated as town, cities, and ranches within a 100-mile radius become deserted due to persistent and severe drought; well defended by its own formidable weapons emplacements; and further protected by the fighter-plane squadrons located nearby at Nellis Air Force Base, our contractual partner in the lease arrangement for the land we occupy. And, most important of all, a little more than a thousand talented children live in the Settlement, the second generation of Hera's kind, who are about to turn eighteen.

Trouble is brewing, although not obviously so as the tale begins. Readers of the earlier volume know that throughout her life, Hera has been obsessed by fears for her own safety, as well as that of her sisters and (more recently) the children. Those threats had always emerged from outside their small circle. Now danger looms from within, from inside the social organization she and her sisters have devised for their tribe, which will be expanding again as the children reach sexual maturity and start to breed. To face it, she will be required to subject to renewed scrutiny the scientific ethos and its meaning for the way in which her tribe should conduct itself in the world, which, in her mind, defines the essence of who they are.

She had not yet faced this question. She would come to understand it as a fundamental challenge to the choices she and her sisters had made, one that must be surmounted; failing to do so would condemn their experiment to failure from the get-go. That question is: What does it mean to actually live out our lives—as a new species of *Homo*—"in accordance with" the understanding of nature that modern science bequeaths to us?

MARCO SUJANA
Director, Facilities & Operations
Yucca Mountain, Nevada

POSTSCRIPT: I have prepared a website where some pictures and maps of the locations described in the following narrative may be found: **www.herasaga.com**

PART ONE

RATBOY'S TALE

VICINITY OF YUCCA MOUNTAIN (SOUTHWESTERN NEVADA)

MARCH 2064

We share 99% of our genes with mice, and we even have the genes that could make a tail.

Jane Rogers (2002)

The Priesthood of Science

For the last few years, I've gotten into the habit of spending a week or so at a time, as often as possible, camping and hiking in the wilderness areas around Yucca Mountain, and occasionally further afield in the deserts of Arizona, California, or southern Utah. Sometimes I'll bring a group of friends from Beatty, but often I travel solo, as a way of taking a quiet break from my hectic routines at the Settlement. Camping alone overnight in the backcountry has become a bit hairy due to the steady increase in the populations of wolves and coyotes, which are happily escorting the few remaining isolated human inhabitants out of the region. Since coyotes enjoy infiltrating human settlements and snacking on pets and other small mammals, we've had to install a secure fence around the entire perimeter at both Yucca and Beatty. But on those occasions when I really want to do an extended solo hike, and Magnus the dog pressures me to take him along, we solicit the help of Alonzo and Quirk, two formerly wild red roan burros (*Equus asinus*) who fear no species of man or beast, no matter how bad its reputation.

Springtime is the best time of year in the American deserts, of course, on account of the flowering plants and perfect temperatures. In all seasons, however, except during the brutal heat of summer, one can also enjoy the other pleasures of the landscape, highlighted against the sky's brilliant blue background: the multicolored layers in the sedimentary rock formations, painted in every conceivable shade of red and yellow ochre, white, brown, and black; the mountain façades, hillsides, and stone mounds, sculpted into a million fantastic shapes by wind erosion; and the simply astonishing variety of plant and animal life sustained in these arid places. I have sat for hours in the perfect stillness of sunrises and sunsets, watching the play of changing light across the horizon.

Of all the nearby places, my favorite is Ash Meadows. Situated in the shadow of the Funeral Mountains in the Amargosa Desert, on the Nevada–California border at the eastern edge of Death Valley, and a mere twenty miles almost directly south of the southern end of our Settlement at Crater Flat, Ash Meadows is a wetland anomaly in the midst of bone-dry desert. Spread over an area of twenty thousand acres, seven major springs and a host of smaller seeps bring ancient "fossil" water flowing to the surface, thanks to a geological fault. Hundreds of bird species frequent the wetlands—raptors, flycatchers, thrushes, owls, sandpipers, herons, hummingbirds, woodpeckers, and more. Groves of mesquite and ash trees ring the major springs, where I usually set up my base camps, most often, however, at Point of Rocks Spring, where water is thrust upward continuously from below, as if a set of powerful pumps had been affixed within the subterranean bedrock. All of the springs are, of course, gathering places for small desert mammals, amphibians, and reptiles, but the vicinity of Point of Rocks Spring often affords a chance to catch a glimpse, too, of the magnificent desert bighorn sheep.

Ratboy's tale was relayed to me shortly after I had introduced myself to him in Ash Meadows, after having shadowed him at a distance and observed his movements with my surveillance gear for the better part of a day. I had invited him back to my campsite and broken out some lunch rations. When, after a short while, he mentioned, almost audibly, the strange nickname he had once acquired, he must have seen a look of astonishment on my face, for he at once leapt to his feet and strode back and forth, his biography gushing forth, unstoppably, the outpouring of an anguished spirit who had spent far too many of the preceding months without any companions other than the desert's elusive wildlife.

"In a way, I wish my father had chosen a woodrat to serve as stepmother and stepfather to me. I often stumble across the middens of *Neotoma lepida*, the desert woodrat, during my wanderings in these dry places, and when sometimes I catch a glimpse of one of these clever little beasts, I never fail to marvel at its tail, fully as long as the rest of its body.

"Their splendid middens—collections of diverse materials laboriously gathered up during foraging expeditions, then bundled together and sprinkled with their urine, which can preserve their treasures for tens of thousands of years—are the reason why the common name of 'pack rat' has been bestowed upon them. Archeologists can identify and date in those middens plant materials and human artifacts revealing the region's older ecology and the ways of the human communities that thrived there eons ago. But I digress.

"Instead my father chose for this purpose another member of the Rodentia order, *Rattus norvegicus*, the brown or Norway rat, perhaps preferring the brown to its close cousin, *Rattus rattus*, the black or ship rat, on account of the former's legendary fecundity and more aggressive nature. True, he might also have avoided the black rat for another reason, since it is scorned and feared as the agent of bubonic plague. But after all, it is not from the rat, nor even from the flea that inhabits it, but from the bacterium ferried along unawares by the flea whence came the Black Death. So in all fairness, should not both the much-despised rat and the lowly insect alike be regarded as innocent parties in this matter?

"Again I digress; it's an unfortunate habit of mine, perhaps an attribute of those who endure a solitary existence and who have no interlocutors to channel their thoughts into ordered ranks. Is it so surprising that my father was celebrated and admired in his native Belize for his cleverness as a scientist? Was he not after all the first to demonstrate to the world that a rat embryo of either sex could be infused with human stem cells and then, having been allowed to gestate and mature, that some entirely human gametes would be expressed in its reproductive organs? Having perfected the technique, did he then not succeed in creating a slew of these dual-species entities—chimeras they are called, I believe—employing in the experiment his own stem cells and those of the beautiful lab assistant he had unsuccessfully lusted after, finally having had to settle for her protoplasm in lieu of her favors?

"Yes, male and female created he them, a large tribe of lovely brown rats bearing inside their bodies a very special genomic cargo. As far as his peers were concerned, this is where the trial had ended; unbeknownst to them, however, he rushed ahead in the hidden facility he had constructed in the slums of Belize City, well away from his regular operations on the Belmopan Campus at the University of Belize. As his rats matured he considered his options. He might have extracted gamete samples from his rodents, screening each batch under the lens of his microscope until he happened upon his prize: The telltale signs of human DNA, some in sperm, some in eggs. The rest would have been easy, involving *in vitro* fertilization and implantation in a human surrogate mother.

"But would it not be far more thrilling to let nature herself craft the finished product? He allowed his rats to breed. Working feverishly yet meticulously with a team of trusted assistants for weeks on end, he sliced into the females' oviducts, one after another, timing his intervention so as to interrupt the embryos at the eight-cell stage of their development, scanning and discarding hundreds until late one night he found in the bloodied mass of tissue under his microscope the prize he had been seeking. Using the techniques perfected for pre-implantation genetic diagnostics, he had first checked the chromosomal numbers—mice have forty, humans forty-six—and then confirmed his results by testing for a set of signature genes of the human type. He had done it! He had shown that, following upon an act of copulation by a pair of his chimeric rodents, a fertilized egg of a very special kind could be gestated in the animal's womb—a zygote that was *fully human*! Again, as it is said, the rest was easy. His blastocyst was surgically transplanted from stepmother rat into the waiting womb of a surrogate human female, hired for the purpose, there to be brought to term. As I entered the world, the occurrence was recorded by the hospital as a 'normal' birth of a common sort, mother present, father unknown.

"My surrogate mother was retained as a maid and mistress in my father's house, both before and after her pregnancy. Having arranged to legally adopt and raise me as his own child, thus permitting himself to monitor daily the developing organism he had manufactured, my father showered me with every advantage in life. Both of my genetic parents were intelligent, lively, and good-looking specimens, and early in life I displayed the same traits in abundance. So far, so good. Since my father was the scion of a wealthy landowning family, I trod easily through my early years, pampered by my many servants, coddled by my teachers, displayed with pride by my father to relatives and friends, and later fawned over by my peers.

"The revelation came much later, one day during my late teenage years, in the course of a furious argument between the two of us over my persistent bad behavior, when my father—or should I call him my maker?—finally blurted out how I had come to be. He boasted to me that he had indeed done all the things I have related to you just now, insisting that I listen attentively to every detail, and hurled his final comment at me while gazing directly into my eyes: 'What I have just told you is no textbook case, you pathetic fool. It is nothing less than your own biography!'

"That I was a spoiled lout I had never denied. Nor did I protest when during our many spats he had autopsied my character using medical terminology, labeling me a pestilential prodigy, a pathogen of exceptional virulence, a Pied Piper of moral corruption leading my contemporaries to ruin and perdition. I recall that, on this particular occasion, I gloated over this description and urged him to escalate his metaphors, taunting him with the challenge that he had yet to plumb the furthest depths of my depravity.

"That's when he struck me across the face and called me 'Ratboy.' He roared with laughter while extolling his own scientific genius, taunting me with gory details of the laboratory routines as I stood facing him, dumbstruck at his account of my genesis. Then his hilarity vanished and a look of dark cruelty passed across his face. 'I made only one mistake, one tiny little error, while scanning your DNA for inherited diseases before implanting the blastocyst in your surrogate mother. Did you know, Ratboy, that some human infants are born with a short vestigial tail that is then surgically excised? Did you know that all humans carry the gene that codes for the development of a tail? Of course, the gene is switched off, but embryonic development is not always perfect. No, how would you know these things, since you have no interest in what you call stupid intellectual games? What I regret now is that I didn't pause at that step to probe into your genome and switch on your tail gene. What fun you could have had! Just think how much more you could have impressed your drunken, doped-up friends by swishing your gorgeous long tail in their faces!'

"I was dumbstruck. What manner of riposte could be made to such news? I remember experiencing no feeling of emotion whatsoever as I proceeded to snatch the heavy golf trophy from his desk nearby and crush his skull with it. A few hours later, I appeared guileless and grief-stricken while urging the authorities to hunt diligently for the gang of thieves that, the evidence suggested, had broken into our house and slain my father in the course of a robbery. Only one concern nagged at me. To discover what he meant when he had blurted out, after the first blow had landed on his head and he saw in my eyes an implacable resolve to finish the job: 'Stop, you idiot! If you kill me they will know how to track you down.'

"It took me only three months to auction off his property and convert my considerable inheritance into liquid assets, which I somehow managed to do despite my state of near frenzy. When shortly thereafter I narrowly avoided capture at the hands of a party of armed thugs, I guessed that something had gone wrong. The next day I managed to lure one of them into an ambush and took him prisoner, discovering through my interrogation that shortly before his death, my father had confided his secret to a university colleague. Undoubtedly, my father's murder had frightened his colleague and he had gone to the police, or to one of the death squads they are affiliated with, urging them to assassinate me. When I eventually confronted and interrogated him, I learned of my father's belief that somewhere in my mitochondrial DNA, there may reside some compelling evidence of my unorthodox parentage.

"After discovering my betrayer's identity and exacting a suitable revenge on him, I vanished into the slums of a metropolis in neighboring Guatemala, where I hatched a plan carefully drawn up to answer but a single question: What am I? I chose a young woman from among the many on offer in the streets and impregnated her, sequestering her in an apartment while awaiting the result. When all the exhaustive medical tests I prescribed for the young infant confirmed his apparent normality, I abandoned mother and child and repeated the experiment. This, too, had the same result.

"Some time thereafter I slipped back into Belize to visit my mother in the house I had purchased for her, but another narrow escape from a band of hired killers persuaded me to flee once more. I hired a small boat to carry me around the Yucatan Peninsula and all the way up the Mexican coast to Matamoros. Then, moving north and west, I skirted the Chihuahuan Desert until I got across the US border and started roaming the deserts of Arizona, California, and Nevada.

"Yet certain difficult questions haunt me still. When I can obtain secure computer access again without compromising my safety, there is an inquiry I must pursue. Having been baptized and raised a Catholic, I shall contact the learned clerics at the Congregation of the Doctrine of the Faith in Rome, explaining my genesis and posing my most urgent query: Do I have a soul? Should the answer come back in the negative, I shall have to consider some other options. The last time I checked, there were a number of souls being offered for sale over the Internet on various electronic trading portals, and the bid price was well within my means."

He stopped just as suddenly as he had begun and bowed his head. We sat together in silence for some minutes. Then he looked up and smiled. "That's a military issue sniper rifle and night vision scope you have. It's a beauty—a Walther, isn't it? May I ask how you came by it?"

"A friend of mine gave it to me. He happened to be a US Air Force general. By the way, my name is Marco." I paused. "I have a computer and satellite communications uplink in my vehicle. After we've finished eating, I'll help you send your message to the Vatican. I'm dying to find out what the official answer is."

I led him back to where my vehicle was hidden under a camouflage net, and after launching the urgent e-mail message concerning the status of his soul to the Holy Office in Rome, I asked him, "What's your real name? I'm certainly not going to call you 'Ratboy.'"

"Jesús Narciso Ramirez. I insisted on using my mother's last name. She's always been a devoutly religious person, and although I wanted to change my given name, she begged me not to do so, and I promised her I would not."

"Well, if I wind up introducing you to my family, the name 'Jesús' might cause a few eyebrows to be raised, so how about we settle on 'Narciso'?"

"My friends used to call me 'Ciso.' By the way, is your family part of that big fenced-off spread to the north of here? I decided to stay well clear of it after I got a look at the electrified fence and abundant warning signs. You folks aren't eager for company, I take it."

"That's an understatement. We picked you up on our surveillance cameras. There aren't many solitary wanderers in these parts anymore. I thought I'd check you out. Anybody who can figure out how to live by his wits in this kind of environment is worth knowing. You don't even appear to have a weapon. How do you eat?"

He pulled out a beautifully crafted slingshot. "Necessity is the mother of invention. I've gotten pretty adept at hunting small mammals with this—and at inflicting enough pain on the top predators out here to keep from being eaten myself. As for the plants, I managed to do some research at an Internet café after I crossed the border from Mexico to the United States, and I investigated the food sources of the ancient nomadic tribes, both what they ate and what they avoided. Mesquite beans, pinyon nuts, cactus and palm fruit, the flowers and seeds of the Joshua Tree and other yuccas; if what's on offer is remotely tasty and nontoxic, I'll feed on it."

He caught me glancing at his deeply scarred wrists. "Yes, I tried to commit suicide, not long after I murdered my father, but my mother found me and managed to get help in time. She was devastated by the prospect of my losing my ticket to heaven. Her anguish saved me, for in response to it I swore an oath to her that I would never again make the attempt. The years I spent running for my life from the squads of assassins toughened me physically and mentally; I acquired the necessary skills in self-defense, and I know how to kill a man with my bare hands before he's even aware of the danger. But I'm not a psychopath, if that's what you're thinking."

"I'd already come to that conclusion. I notice that you're not packing much gear. When did you arrive in the Mojave?"

"About three weeks ago. I spent the winter in northern Mexico and began to move north as the weather warmed." He displayed the contents of his kit: goose-down sleeping bag, blanket, folding shovel, a second pair of rugged sandals, socks, gloves, scarves, hat, three large canteens, waterproof matches, several knives and a sharpening stone, one change of clothing, medical supplies, moneybelt, small binoculars, a few family mementos. And a small bag of steel balls. Noticing my interest, he remarked, "Those come in handy when I need to do some real damage to a predator—man or wolf—with my slingshot."

As we relaxed after lunch he insisted on my matching his tale with some suitable piece out of my own biography, so I related the highlights of my aborted interplanetary trip.

"This will sound just a little crazy, but the expedition was to take us to Mars, a somewhat inhospitable planet from the standpoint of mammals from earth. The atmosphere is poisonous, 95 percent carbon dioxide; the average annual temperature is –53 degrees Centigrade and it can go as low as –128 degrees Centigrade, which on the Fahrenheit scale is about 200 degrees below zero. Gravity on Mars is one-third that of earth's, and the atmospheric pressure is about one percent of what we're used to. In a nutshell, nobody's going to fling off his or her spacesuit and roll around in the sand to celebrate a safe landing. On the other hand, as we know from the many cameras we've planted on the surface over the years, at least in the intervals between dust storms, the nighttime view into space is truly astonishing. There's hardly any atmosphere to speak of, so the sky confronts you in stunning clarity, and you gaze on Mars's two small moons as well as a sparkling canopy of starlight—and planet earth, of course.

"The basic objective for our mission was to set up a secure facility in which scientific experiments, surface exploration and sampling, and continuous environmental monitoring could be carried on for decades after we left. Every piece of equipment was designed to be self-repairing and to be under remote control by radio signal from earth."

"Amazing. How many people were on the trip?"

"Thirty of us trained for the mission, men and women, of course, including a rotating skeletal crew who would stay on the orbiting space vehicles. Our trip was timed to arrive in the northern hemisphere during its summer season because the relatively colder temperatures there would be far preferable to the worst of the dust storms, which occur in the southern hemisphere at summertime. Our mission was a simple one: We had up to six months in which to construct and equip—using prefabricated materials—a suite of underground living quarters, complete with laboratories and storage bays for the robotic equipment that would continue to roam the surface after we left. Part of the mission was to see what eighteen months away from earth's gravity would do to our bodies.

"We pretty much knew what to expect as a result of studies on previous astronauts. Cardiovascular function can be maintained with exercise and physical activity, and we had plenty of that. The chief risk is bone density loss, especially in the lower half of the body, as a result of both the zero gravity of space and the reduced gravitational pull on Mars, because the skeleton isn't supporting its weight. The calcium imbalance resulting from loss of bone density presents a serious health problem, as is the increased risk of fracture. Some drug remedies have been developed to protect against these effects; we were part of the ongoing experiments to see which ones work best."

"So," he interrupted, "how did you like living there, if I may ask?"

"We never went. The whole thing was called off at the last minute. I called it off—from outer space."

He whistled. "That must have been an expensive call, no?"

"You don't know the half of it. Some twenty-seven billion euros of the sponsors' money had been invested in the construction of the spaceships and the supplies for the underground city. Not to mention the countless billions the US government had generously donated to the preparatory work over the years."

"Did you say 'sponsors'?"

"Yeah, this whole thing was one of those public-private partnerships, which included the Vandenberg Air Force

Base in California. The private end of it was a bunch of insanely wealthy families from the international criminal set. It sounds like a lot of money, I'll admit, but I suspect it was only chump change to them."

"Just exactly where were you when you called it off?"

"I was floating around in the International Space Station as part of the team sent up in advance to do a final check on the preparations. The spaceships actually depart from there; they're nuclear powered, which makes for a fast trip to the Red Planet. My title was commander of technical operations. I had the responsibility to make the final call—go or no-go. I was dying to go!"

"So what happened?"

"I discovered at the last minute that some senior managers at Vandenberg had been covering up the failure of one of the most important experiments—dealing with protection of the astronauts from cosmic rays. Without adequate shielding, something like one-third of your DNA will be cut to pieces for every year you spend in space. This was known for a long time, of course, so they tried everything they could think of to provide the space vehicle's shielding—thick shells, elaborate magnetic shields, even gigantic electrostatic shields to deflect the incoming cosmic rays. They opted for the shell, claiming to have found a nanoparticle matrix that would do the job. It didn't. But that little fact came out only after I had gone up there."

"Why would they lie about it?"

"Figure it out. There was a humungous investment in the project, not to mention all the surrounding hype. So they made a simple risk calculus: The astronauts would be okay for the duration of the trip there and back, plus the time spent on Mars. What we didn't know wouldn't hurt us—not at the time, anyway. You don't feel the cosmic radiation itself as it's striking you. Only later would the cancers develop, by which time the mission would have been accomplished. Pretty clever, wouldn't you say?"

"How did you find out?"

"One of the technicians at the Station who had just been diagnosed with bone cancer blew the whistle. When I told the ground personnel I was calling off the mission, and when I resisted all their suggestions as to how to 'work around' the problem, I became the problem. So they tried to have me arrested on trumped-up charges of sabotage, except that no one could figure out how to actually put the handcuffs on. When I was assigned to do a spacewalk outside the Station, it occurred to me that I might have an 'accident'. Do you remember that scene in *2001: A Space Odyssey*? Anyway, I had some very influential friends at Vandenberg, so they put an end to this nonsense and called the whole thing off.

"Frankly, Narciso, I've never thought much of the scenario in which Mars figures as a home away from home, sort of an extraterrestrial resort, for humans. The idea that humans and the other living creatures they might bring along from planet earth for company, such as pets and plants, eventually would adapt, in an evolutionary sense, to the starkly different features of that environment, is intellectually interesting, to be sure, as long as the emigrants remember just how much time that process would take.

"There's a cute story in one of the old volumes compiled by mission-to-Mars enthusiasts, in which the well-adapted permanent residents on the Red Planet weep bitter tears as, watching from afar, they see a gigantic asteroid smash Earth to smithereens. It seems not to have occurred to these folks that it's all a matter of probabilities, and it's just as likely that Earth inhabitants would be the ones doing the watching, while lamenting the wasted effort on Mother Nature's part in redesigning her products for life on Mars."

We remained silent for some time as a group of desert bighorn sheep moved across our field of vision. After they disappeared over the crest of a hill, I turned to Ciso. "There aren't many Walther sniper rifles in circulation. You must have some experience with this type of gun."

"I do. It came in handy when the assassins were stalking me in Belize."

"I've got a second gun, not quite as good as the Walther, in my truck. And I set up a little firing range not far from here. Let's go do some target practice."

"Then these are perhaps the end of days as foretold in our Bible, aren't they, Hera?"

She was sitting on the veranda of her suite at the edge of Crater Flat, beneath the brow of Yucca Mountain. With her was Jacob Hofer, senior minister in the group of Hutterian Brethren colonies located at her Yucca Settlement. They had been conducting a brief review of the latest eruptions of violence and chaos occurring outside the borders of their little desert enclave.

The bond of friendship between Hera and Hofer had begun early in their association and strengthened during the long conversations they held monthly. Ostensibly these were designed to review administrative matters in the relation between the Brethren and the Settlement, such as financial issues, crop planning, security concerns, water management—a constant worry in the bone-dry desert they inhabited; also the welfare of their farm animals, and overseeing the Hutterite youth who had been allowed to work with Settlement personnel outside the colonies' gated areas. As the chief administrative officer for the Yucca territory, I was required to attend, reporting on earlier action items and recording new ones. Once these matters were disposed of, both enjoyed turning their conversations to what they liked to call "spiritual affairs," and usually I lingered and listened.

"As you know, we believers do not fear the end of days," he continued. "These events are foretold in our Bible, so we must accept the truth that this epoch will come to pass. Our hope for eternal salvation is rooted in the events that are set in motion then, for they signal to us that God is ready to bring to a close man's earthly existence. He is ready to assemble the heavenly host for Judgment Day." He turned toward her. "You are not such a believer, as I surmise. I wonder what comforts you as you contemplate this prospect."

"I, too, must accept the truth that one day my species will become extinct everywhere on earth," she replied after a pause, "for it seems that nature dictates this fate to all species. I don't think I'm necessarily required to regard such an event as a cause for celebration. But as far as my own personal case is concerned, I wouldn't try to deceive you, Jacob. No self-conscious being can face the prospect of her own death with a completely untroubled mind. I, too, need my consolations. I derive them from different sources, that's all."

"Can you tell me what they are?"

"Chief among them is my music. I'll offer you just one example now, which includes a story about the work that haunts me constantly—Mahler's *The Song of the Earth,* a set of six pieces for orchestra and solo voice. The last of them, entitled "Farewell," is fully as long as the entire five that precede it. It opens with two chords in the low register, a kind of ominous chime, which the composer also inserted again to mark its mid-point. At the place in the score where the tam-tam is played, Mahler inscribed a single word: *Grabgeläute.* You're fluent in German, Jacob. How would you translate it?"

"One might best express the English equivalent as 'death knell.' But the connotations are a bit more sinister, if you will. 'Grab' means 'the grave,' so one might render this compound word more literally and expansively as 'the peal from the grave.'"

She nodded as he went on. "I don't know this composition, so I fail to understand the consolation you find in it."

I interjected. "I recall you told me once, mother, that a Russian composer—I think it was Shostakovich—had remarked that if he were told he had but a half-hour to live, he would want to spend it listening to this Mahler song."

"Indeed I did. For Mahler had composed his masterpiece under the shadow of death. He began it in 1907 shortly after his beloved eldest daughter, Maria, had succumbed to scarlet fever and diphtheria. Mere days after his daughter passed away, Mahler discovered that he had a serious heart disease. He had been so closely attached to Maria that when his own fatal illness gripped him a few years later, he told his wife that he wished to be buried in Vienna in the very same grave where his daughter's remains lay. Like the aged Rembrandt, whose late self-portraits record mercilessly the process of fleshly decay, Mahler had the moral courage to force himself to face squarely the prospect of his own death in his art."

Jacob interrupted. "You've told me previously how much you are affected by the music that is based on the Christian service for the dead. Is this a requiem, then?"

"No, and that's one of its remarkable aspects. You asked me where the consolation in it is. After the second sounding of the tam-tam, its long conclusion begins. The figure of death—"Freund Hain" in German folklore—appears in the lyrics but is then displaced as the music is drawn to a close by another vision entirely, a vision of the eternity of our 'beautiful earth,' our only home. Mahler based the rest of his lyrics on ancient Chinese poems, but he wrote the concluding stanza himself. It ends with a single word: 'Forever.' The English composer Benjamin

Britten remarked that this final chord is 'imprinted on the atmosphere.'

"The difference will be obvious to you, Jacob. The requiem mass or service for the dead, whether Protestant or Catholic, always ends with the hope of the resurrection of the soul. So Mahler's work is, if you will allow me to say so, a kind of anti-requiem. Here it is not the individual soul—the residue of the human person—that endures, but rather the earth itself, from which we all spring and to which we return. However much we might long for eternal life, I believe we are not entitled to sustain such a belief. We call ourselves *Homo sapiens* on account of our reason, and our reason presents us with irrefutable evidence that all beings born of the earth return to it forever. There are no exceptions."

"You will forgive me if I say to you that this strikes me as meager consolation indeed."

"Undoubtedly. But to my way of thinking this is all that is permitted for us."

After a short hiatus he remarked, "Let's leave aside the 'end of days' idea theme for now, Hera, and return to what was said when we were taking leave of each other after our last conversation. You are apparently intrigued by the continuing effort to 'reconcile' religious belief with modern science. Why is that of such interest to you?"

"Perhaps it's the sheer implausibility of the whole enterprise that makes it so irresistible. We've spoken about similar things before, and you know I always preface such remarks with the proviso that, for me, organized religious practice—based on a specific doctrine of faith—isn't the same as the search for meaning in existence."

"Yes, I remember. We Hutterites have a core doctrine of faith, as you undoubtedly know, based on our Apostles' Creed, and I suspect there has never been any type of religion, properly speaking, where this element was entirely absent."

"Some of those who are obsessed with the reconciliation of science and religion are, of course, traditional churchgoers. But whether they are or not, from my standpoint all of them are right to be worried about the future of faith-based religion."

"But why? Christianity is two millennia old, older still if you add its Judaic roots. Why should we believers be anxious in the presence of science, a relative newcomer to the scene? For me, it's an exceedingly simple matter: Nothing that happens in this world can undermine my faith. Like everything else, science is an affair of mankind. My faith comes from God. The two never meet, so to speak. What is there to reconcile?"

"That's a good question. You will concede, I hope, that the brethren in your communities are cut off from the secular world to a far greater degree than are the mass of modern religious believers. In that world, modern science has come to play a role far more influential than scientists themselves would have imagined in earlier times. By the close of the preceding century, science was in the news continuously, mostly in terms of medical research and new therapies, but actually broader still, extending even into the mysteries of theoretical physics. Why, even you might have heard of the so-called 'God particle,' which was the subject of much coverage in the mass media."

He laughed merrily. "No, I can't say that I have. What could this possibly be?"

She laughed, too, in response to his surprise. "I thought the phrase would get a reaction from you. The reference was to a scientific hypothesis of the time, concerning a subatomic particle called..." She paused. "Oh dear, the name is...."

"The Higgs boson, if I'm not mistaken," I interjected

helpfully. "You told me about it once, on one of those many occasions when you were filling my head with useful information."

"Yes, thank you, Marco. Do you see why I need you to attend these sessions? The Higgs boson was, at that time, pure conjecture; it had never been observed, but physicists had a profound need to suppose that this tiny and elusive entity might exist for a very simple reason: They required *something* in the nebulous world of subatomic particles that would account for the fact that matter has mass! You see, science has always prided itself on its advance over older metaphysics, which was bedevilled by the question, 'Why does anything exist, rather than nothing?' Scientists don't much bother themselves with the 'why' aspect of that question, but they do worry when their main explanatory model has a big hole in it."

Hofer's confident rejoinder showed how pleased he was that their exchanges had arrived at this point. "There's a simple answer to the 'why' question: These things exist, quite simply, because God willed them to exist. What other explanation is needed?"

"So you believe, and I can accept your explanation in those terms. But it won't do for scientists. In the modern science of nature, every proposition about the characteristics of the natural world must be capable of being experimentally verified. If a new proposition represents a conjecture derived through a process of reasoning, say, a set of equations, then it may be accepted provisionally; ultimately, however, it must generate predictions about the behavior of matter and energy that can be tested."

"For us it would be blasphemous to think that God would be required to subject Himself to an experiment designed to prove His existence," he commented.

"In a way you are making my point for me," Hera replied. "Science chooses to limit itself to propositions about matter and energy that can be tested in that way. Religion has a completely different type of explanation for some of the same phenomena. At first glance it's hard to understand why anybody could be bothered to try to reconcile the two. Sometimes the whole business just seems silly. Shortly after the 'Big Bang' theory of the origins of the universe gained popular notice, about a hundred years ago, the Catholic Pope at the time concluded that this theory confirmed God's existence. That statement was absurd on its face. A subsidiary issue was his failure to demonstrate that it was *his* God—the Pope's God—whose existence had been so conveniently confirmed, and not some other equally well-qualified candidate, say, the Aztec creator-god Huitzilopochtli."

Jacob smiled. "Perhaps that's why I and my brethren don't concern ourselves with these sorts of conjectures, Hera. Our faith suffices. But the matter does interest me. Why do you think some have felt the need for such a reconciliation?"

"I think it's no accident that this interest peaked at the time when changes in society had thrust science into the public's face, so to speak. Toward the end of the twentieth century there was a growing awareness of science's impact on daily life, especially on account of genetics. All of a sudden, science had a kind of immediate 'presence' for everyone; it had become personal. Earlier, many things of great importance to ordinary people, such as good health, recovery from disease, having children, and longevity, had been the subject of prayer. Then, gradually, science began demonstrating to everyone that there was another plausible set of reasons for all the things that happened in people's lives. In other words, science was no longer only the progenitor of clever electronic gadgets; now it started to earn respect and recognition for its views on issues that affect people at the deepest levels of personal well being.

For example, they learned that scientists were hunting for what the popular press dubbed the immortality gene. Curiously enough, nobody appeared to be overly upset about the prospect."

"So science was beginning to penetrate more deeply into the core regions where religions had always had the most authoritative voice."

"Indeed. Some could hardly fail to notice the implicit challenge: Who needs faith once science call tell us everything we need to know about how to control life?"

"But not how to guide life toward fulfilling our sacred duty, Hera; not how to be a good person, nor what must be done for the salvation of one's soul in preparation for life after death."

"Admittedly, no. But would you allow me to press you on those very points, Jacob, my dear friend? I know these are sensitive matters. Will our friendship be able to endure this interrogation?"

"You need have no fear on that score. Like many of my brethren, I hold that life tests our faith every day, sometimes severely so."

"Good. So far, I have not raised what is, as I see the matter, one of modern science's most powerful thrusts against religious faith, one that is rarely mentioned in the rather sentimental 'feel-good' literature about reconciling science and religion—namely, it's quality of universality. And here I mean that term to be understood quite literally. What are called the 'laws of nature' hold everywhere and across all time, at least in the dimension of the universe we happen to inhabit, not just in the here and now on this puny, insignificant hunk of rock orbiting a relatively small star. No other aspect of its worldview so sets it apart from the sphere of religious faith as does this one."

"I beg to disagree! Nothing is more universal than God's majesty and power. After all, He created that puny universe of which you speak, and is the author of those other mighty domains—heaven and hell—as well."

"Thus we come immediately to the difficult part. For in fact there is no *universal* faith as such, no religion as such, to set against the scope of science's vision. Rather, there are many particular faiths across the face of the globe, as well as many, many more that once flourished and then collapsed, never to rise again. Literally hundreds of systems of faith, peopled by thousands of gods and goddesses, stretching back in time a full seventy thousand years to the first worshippers of the python deity unearthed a few decades ago in the Kalahari Desert. What unites all of them is only the simple fact that none has ever commanded the allegiance of more than a fraction of the earth's human population at any one time."

"Hera, my faith is utterly unaffected by the knowledge I possess about others' beliefs. That is not one of the tests I mentioned a moment ago, by the way. My struggle—and that of the community of which I am a part—is to remain worthy in the eyes of the God we worship. We are not called upon to find fault with the sincerely held creeds of other communities of believers."

"Jacob, please! The revered founder of your sect, a pious and devout man, as I believe, was immolated at the behest of those representing a different faith! Jacob Hutter's followers, your forebears, were hounded from country to country by people who claimed to be acting in the name of God, more precisely, your very own God, the Christian God. Some of them were forced to re-convert to Catholicism on pain of death! How can you maintain your attitude of indifference to the conflict of faiths in the light of your own terrible history?"

"And for us to be worthy in the eyes of our God, we are commanded to forgive them for their sins, as you know. We do not necessarily know what motivates such deeds, nor do we assert the right to judge them, for that, too, is God's business. But we will be held accountable for how we ourselves act, including what actions we take in response to the blows inflicted upon us. We are pacifists, Hera. You may know the story of the four brave Hutterite boys who, despite our protests, were drafted into military service in this country during the First World War, and who were so brutally beaten on account of their non-cooperation that two of them died. They suffered these unjust punishments without retaliating."

Hera reached across to his chair and touched his arm. "Jacob, in my mind you and your people are the truest, most authentic of Christians, and I admire you with all my heart. I would not presume either to test or to judge you, as I think you know. In raising these issues with you my primary aim is to clarify my own way of thinking and belief, to hone it so that it can continue to serve me well in the challenges that I myself face. You do understand, don't you?"

"You must not worry. Just go on."

"The issue I wish to confront pertains to what I take to be religion's self-proclaimed strength, namely, the moral system lying at the core of every system of faith. This is seen to be, even by the advocates of secularism, its trump card, in a way. In a word, the assertion is that religion helps to make people good, assisting them to reject their natural urges to do harm to others; in this respect, faith and belief are widely regarded as being among the main pillars of what we call civilization."

"With the proviso that we are commanded by God to act in this way. Certainly I believe that our earthly life is made better thereby, although strictly speaking that may be an incidental benefit, so to speak."

"So in light of what has just been said, you would undoubtedly reject a charge that most religions contain a nihilistic core. Before you reply, let me explain how I am using that term. For me, nihilism is the rejection of the idea that there are objectively true systems of ethical values, an idea developed by many philosophers as well as theologians. Your Bible's Ten Commandments are one example. So is the philosopher Kant's categorical imperative, which directs each one, as an individual, to act in a way that would be seen to be just by every rational being. Nihilism responds by saying that these are just arbitrary preferences."

"I understand and agree with your view of nihilism, but I am entirely lost in the rest of your account. You concede that most religions, including mine, promulgate systems of universal values. Thus, how can you possibly say they are fatally infected with what they oppose?"

"The solution to this apparent paradox is straightforward because in most cases, at least, the allegedly universal moral law stops at the borders where different religions meet. There, at that interface, a metaphysical black hole opens, revealing a dead zone of the spirit. Locked there in a vise-grip permitting no escape, the moral law implodes, collapsing into itself like a dying star, to be torn apart in periodic orgies of murder and mayhem. Meanwhile, on the fringe of this ominous singularity, the watching, cheering hordes of the faithful gorge themselves on a deadly mixture of hatred and innocent righteousness."

"Please, Hera, you seem to be carried away with the force of your own thoughts. My faith allows no such hatred to be expressed against others, whether they be believers or not."

"And I say back to you, please, Jacob, look beyond your own precious faith. Yes, you are Protestants, but you are also Christian pacifists. Over the centuries your people have paid a high price for their peculiar interpretation of their faith, as you yourself conceded earlier. You follow the

Gospels' injunction to 'turn the other cheek.' But I ask you, how many other believers do so?"

"Again, I can only say that we are answerable to God, as those others are. He is the Judge and He will hold them to account for what they do."

She smiled again. "How I admire your steadiness and peace of mind. How often have I wished I could appropriate your inner clarity for myself! Yet as much as I desire it I cannot go there. Instead my mind fills with images of Sunnis slaughtering Shiites inside their mosques. They worship the very same God—the God each appeals to for assistance in exterminating the devout as they kneel in prayer! Meanwhile, sitting uneasily on his throne in Rome, His Holiness hopes that no one is so crude as to remind him of the centuries of torture and assassination his church unleashed against dissident Catholics, Protestants, and Jews alike. The justification was brutally simple: Religious tolerance breeds heresy.

"May we exempt the peace-loving Jews? Not all of them, certainly not the most orthodox who recall fondly the 'good old days' recorded in the Old Testament, when Jahweh regularly exacted vengeance against the women, children, and men in neighboring tribes whom He labeled foul idolaters. What was their error? They thought they were entitled to worship a deity of their own choosing!"

She stopped. "I'm so sorry, Jacob, you're right; I get started on a rant and can't find the stop button." She looked at him. "But, really, when I read statements such as those by some Muslim clerics who refer to Jews as 'the sons of pigs and monkeys,' what else can one say except: 'No, you're quite wrong there, both pigs and monkeys are intelligent mammals, unlike you.' Now I *will* stop, I promise."

After a long pause he spoke again. "All members of the Hutterian Brethren communities learn their own history, Hera. How could I deny what you say happened to us? At the same time, I must say to you that the evils others do cannot possibly undermine my own grounds of belief, even if they invoke the name of my own God to justify their acts. Your metaphor of the black hole where religions meet is quite striking, I concede; I would expect no less of you, based on the long record of our conversations. But is it fair to imply that it is unchanging? Isn't it true that within Christianity, at least, there is a greater spirit of ecumenicalism than there used to be?"

"Not a great deal, as far as I can tell. Let me ask you this. Suppose you had a visitor one day, perhaps a suave, beautifully robed cardinal from the Holy Office for the Propagation of the Faith in Rome. Imagine that he said to you: 'Those bad days of old are over and will never return. We admit that we once carried out forced conversions, requiring adherents of other faiths to join our Church upon pain of death. We promise never to do this again.' Would you believe him? Would you bet your life and soul on his word?"

He laughed. "I would not bet my soul on any earthly wager, no. Then again, I'm not in the habit of risking my soul, except on the promise of God's beneficence."

"As I have long thought, you are a wise man. Still, it seems to me that a shadowy doctrine lurks deep within all monotheistic faiths, especially those where a militant deity inspires similar behavior in his followers. This doctrine is well expressed in the Latin phrase that originates in ancient Catholic texts and papal pronouncements—*extra ecclesiam nulla salus,* meaning there is no salvation outside the church. Fair enough, but which one?"

"I can only answer for my own faith. To profess to be a Christian means that you must accept the truth of the divinity of Jesus Christ. This is, as you may know, the 'justification by faith alone' in the name of which Martin Luther launched the Reformation. All who sincerely do accept that truth may hope for salvation, forgiveness of sins, and eternal life, whatever form of Christian doctrine to which he or she adheres."

"Fair enough, but as you well know, even in our present century, a full five hundred years after the Reformation, Pope Benedict XVI went on record as reminding all Protestants that, unfortunately, their religions are not true churches—thus confirming, in effect, the doctrine that no matter how sincere their faith, they cannot hope for salvation. And what of the devout Muslims, who most certainly and emphatically renounce the Christian truth? And the equally devout Jews? Shintoists? Buddhists? Hindus? Shall I go on?"

"That is a harder question, I must admit. I suppose what I'd prefer to believe is that my God, being infinitely compassionate, just, and wise, would look kindly upon anyone who had led a life devoted to seeking goodness and shunning evil. But in the end I don't know what might be the outcome of His judgment, in terms of the afterlife, for such a person. I affirm again, that is God's business, not mine. I am obliged to care for the condition of my own soul and, to the extent I can, the souls of the community of my brethren."

After a short silence between them, he spoke again. "May I turn the tables on you for a moment, my friend? May I turn the interrogation in your direction?"

She laughed heartily. "Of course, that would be eminently fair after what I have subjected you to today."

"You are an atheist, if I'm not mistaken."

"I prefer to be called an agnostic, Jacob. As you are aware, I accept the scientific worldview. This obliges me to seek confirmation for whatever beliefs I hold in evidence, deduction, and experimentation. From such a standpoint, I affirm only that no good case has been made regarding the existence of a nonmaterial deity. None, at least, that attains the standard of proof I expect in every other matter before my mind. Or with respect to how I should conduct myself."

"Fine, an agnostic, then. It actually makes no difference to me. I'm glad you raised the issue of conduct, however. Is it not true that every code of human conduct has been based on some sort of belief in higher powers? Haven't you neglected this truth, earlier in our session, in focusing on that gulf between different forms of faith—what you called a 'metaphysical black hole,' if I remember? You focus on the special case, and ignore the general rule, which is, simply speaking, that the moral law is a product of religion, God's gift to us. People violate the law often enough, I concede, but in its absence, what would save us from endless brutality and murder?"

"A fair question. Let me just preface my answer by saying that for me as well as for you, the moral law forms the bedrock of civilization. In its absence, nothing else we cherish, neither justice nor culture, nor any of the comforts of the body and the joys of the mind, would be given to us."

"Exactly. Unlike you, I claim that the bedrock itself rests on a deeper foundation, namely, God's love for His people. Our hope of living in a state of righteousness originates there. If you dispense with this truly unshakable edifice, the moral law will soon be swept away in the maelstrom. What say you? What do you offer in its place?"

"Nature."

He seemed to be genuinely shocked by Hera's quick riposte. "What could you possibly mean by that answer? The inclinations originating in our nature are exactly what God's moral law is designed to protect us against."

"I apologize. My curt response must have appeared flippant to you, which was not at all my intent. So, phrased more fully, those aspects of nature revealed to us by the modern science of evolutionary biology are what furnish the moral law with its unshakable foundation. From this perspective the causal sequence is reversed: Most settled human societies do feature ethical systems grounded in religious belief; in point of fact, these are authentic products of our nature, not a counterweight hung onto it. The instinct of religion, if I may put it thus, is grounded in the neurobiology of the higher mammals."

"You will see why I reacted as I did, since you appear to be turning what I and my co-religionists believe on its head. Nevertheless, I confess I am intrigued and wish to hear more. I also strongly suspect you are prepared to offer a longer disquisition for my benefit." There was a bemused expression on his face.

"Oh dear, you know me all too well! Do I victimize you with my discourses?"

I couldn't resist. Turning toward Jacob, I said, "At least you should know you are not alone in suffering this affliction. Having had far more experience than you in this matter, I can testify that, after passing a certain threshold of exposure, the victim is gradually overcome with the delusion that the treatment is actually quite beneficial."

They both laughed and Jacob waited for the next round.

"Some ancient philosophers argued long ago that morality was rooted in instinct, but this was merely a speculative idea, unconvincing to most. The huge advances in neurosciences at the end of the twentieth century opened an entirely new approach. Scientists identified the neural circuitry in our brains that provide the natural basis for our ethical behavior. And they found its evolutionary origins in the brains of our primate ancestors, monkeys and chimpanzees."

"We're talking about the difference between right and wrong, I assume? And you say that people are born with the ability to distinguish the two?"

"Yes, and in that sense it's instinctive."

"But we and others teach our children what's right and what's wrong from an early age. How can you be so sure about cause and effect?"

"Well, you must understand that the conclusion I'm reporting here is based on many hundreds of experiments and, I suppose, by now, many thousands of published scientific papers. I've only read the general summaries, of course, but let me say that I find the explanation and the evidence to be convincing. Remember, you are interrogating me about my state of belief, as it were; I'm not in a position to persuade you to accept it, nor is that my purpose."

"Mine seems simple by comparison: Moses received the Ten Commandments on Mount Sinai. Most versions of Christianity then developed more elaborate ethical codes derived from that set of God-given directives. Can you give me a comparable version from your science?"

"Indeed I can, acknowledging the fact that we are both simplifying. I can express it in the form of three operational capacities, and one more general faculty of the mind on which those three depend. They are functional aspects of the neural tissue that makes up quite specific regions within a human brain. In this sense they are just like the language function: All humans are born with the innate ability to learn the exceedingly complex grammar of languages.

"The first capacity is called 'theory of mind.' It means that we can imagine ourselves in another's place. We assume that others are thinking and feeling beings much like we are."

"Is this the same as the sense of empathy?" I asked.

"Yes, in a way, thank you, Marco. One key aspect of the theory of mind is that it allows us to make inferences about others' motives. In a very practical sense, this capacity is indispensable for moral thinking. It means we can distinguish between motives and actions, that is, about the *intentions* of others when they do certain things. Whether they did them deliberately or accidentally, for example, or whether their motives were altruistic or selfish."

"Of course, intention is the very core of moral behavior, as well as the criminal law," Jacob commented, "although I must say we've stumbled into murky realms here, Hera, because some will go to great lengths to disguise their true motives. We would say that sometimes only God can peer deeply enough into the human soul to discern the truth of the matter."

She chuckled. "True enough. Believe it or not, psychologists have amassed a huge list of publications on the subject of cheaters. For social animals such as humans, who are highly dependent on cooperative ventures, there's a huge evolutionary advantage to being able to detect cheating, which violates the dictum of fairness. Did you know that quite young children do very well on tests designed to ferret out such behavior?"

"No, I confess I did not. You will not be surprised when I maintain that only God can do so infallibly. But please, continue with your list."

"The second capacity is the sense of fairness. This is interpreted very broadly as a feeling of the 'rightness' of, say, the allocation of resources among different individuals. In philosophical terms, it's what we call distributive justice.

"The third capacity is altruism, a desire and willingness to help others."

Jacob intervened. "And you are claiming that we are born with these features?"

"Yes. According to the researchers, all of them are clearly in evidence among human children at an early age. In the case of altruism, it has been demonstrated at the age of eighteen months, that is, in the period when language itself is just beginning to develop. The others I mentioned arise by about the age of four."

"So that is the basis for your assertion that these capacities are inborn rather than being taught. You can imagine, I'm sure, that it's easy for me to accept this premise, simply because I believe strongly that God designed man so as to be able to distinguish between right and wrong. That is the basis of each person's responsibility for his actions and for the divine justice that is meted out to everyone at the end of time."

"Of course. However, from a scientific perspective, a different type of evidence is sought. Since evolution and comparative genetics shows that we are a late branching of the primate line, from chimpanzees in particular, scientists seek additional evidence in primate research to illustrate the point that these capacities actually are outcomes of the evolution of species. And they have found it: Altruism, for example, is revealed in behavioral studies of very young chimps, albeit in a less robust form."

"This is all very interesting. If I remember your opening words, there is one more piece to the puzzle."

"And that is a general faculty of our minds that underpins the three capacities—namely, a sense of selfhood or 'agency,' the sense each of us has that there is an 'I' who is the continuing author of 'my' acts, intentions, and goals. We all have a very strong *feeling* of ownership of this entity we call our 'self', and, of course, we attribute such a feeling to others. On this basis, and using the capacities described above, we are able to judge the worthiness of our own intentions and actions, and to do the same with respect to others. We create and share a set of moral principles using these features of our minds."

"That sounds very much like what we would call the 'will,' Hera. Recognizing the will is a way of accepting responsibility for our actions and of expecting others to do the same."

"Precisely. Here I want to give you just one more example from the results of primate research, showing that this, too, arose in the evolution of species. A remarkable chimp named Sarah was the subject of behavioral experiments over many years, and in the course of that experience she grew to like certain trainers and dislike others. In one test she was first asked to watch a videotape in which a trainer attempts to use a stick to reach food. Then she was given a set of envelopes each containing three pictures, each set showing a different trainer, known to her, in the act of using the stick. One picture showed the act being performed successfully; the second, unsuccessfully; in the third, the act shown was irrelevant to the task at hand. And in every instance, Sarah chose the picture indicating success for the trainers she liked, and one of the other two pictures for those she disliked!"

"What did the researchers themselves conclude from the experiment?" I inquired.

"As I recall, they deduced from Sarah's choices that she was aware of her own goals and intentions, and could discriminate between her own and the goals of others. Also, she had the capacity to make implicit judgments about the 'rightness' of the actions of others in her own terms."

"It's clear to me now," Jacob remarked, "why you believe that natural evolution supplies a good explanation, entirely by itself, for the emergence of what could be called an ethical or moral 'disposition' in our species. You have told me many times about the close genetic affinity between us and the chimpanzees. If those traits you called capacities arose in primates, most fully in chimpanzees, then by logical extension humans could possess them in a more highly developed form."

"Indeed that is what I believe. Remember that this genetic inheritance may be regarded as a set of *predispositions* we are born with. As an ensemble of capacities that are present in the neurological attributes of human brains, they still must be nurtured, reinforced, and exercised by each of us. Culture and education play an indispensable role in this regard, as do our civil and criminal laws. As do religions, too, at least in their better moments, when they forget about their recurring need to consign unbelievers to hell simply because they profess a different faith."

"Let's not dwell again on your last statement, please. I find myself quite comfortable with this account of the evolutionary perspective, for the simple reason that you and I wind up at a similar endpoint after starting with radically different premises. We both affirm the centrality of the moral disposition in the lives of our people. It would be easy for me to believe that God designed a clear path, winding its way through the evolution of species, toward this goal, a path that terminates in the emergence of mankind. I hasten to add, by the way, that my faith doesn't require me to believe this story, only that its ultimate outcome was a necessary part of His design, and not a contingent or accidental occurrence."

"True enough. And that is the point where our beliefs diverge. For me, there is no design or purpose imposed on the natural order, and no need to presuppose the contrary in order to arrive at a complete understanding of the emergence of our species on earth. But when I am with you, I am much more inclined to emphasize what we share than what divides us."

"It's all well and good to focus on what we share in our beliefs, but I don't want to get quite so comfortable just yet. Earlier you spoke eloquently in pinpointing what you call a hidden mass of nihilism at religion's core, which is, frankly, an idea I still find shocking, although I promise to examine it further. So let me try to turn the tables on you, if I may. For you, modern science is the grand alternative to our faith, and yet, in subscribing to it so wholeheartedly with your mind and soul, are you sure you have evaded the corrosive nihilism you appear to fear so much?"

Hera fell silent for what seemed like an eternity.

"Have I hit a nerve?" Jacob asked.

Finally she replied, "I think I know what you're getting at, but why don't you explain a bit more fully what you have in mind?"

"We've had dozens of these conversations over the years. I find the intensity of your faith in science—if I may put the matter thus—quite touching. I, too, can marvel when you read me excerpts from the great book of new knowledge and new powers that science delivers to mankind. On the other hand, in my faith, the grant of powers from God, from whom we received dominion over the earth, arrived hand-in-hand with a system of moral law. I see no such codicil in the case of your science. Let's not be naïve: Knowledge and power begets temptation and, when it is unresisted, temptation begets evil. This aspect of your faith troubles me."

She sat silently again for awhile. "All right, I accept the challenge. Here's my answer. In your faith, power and the moral law have a common source in God's dispensation. We humans are the passive recipients of both. Whereas in my faith, to use your terminology, the two are clearly disaggregated. Modern science is a product of human reason, but, in a way, so is modern society. The exercise of reason is their common source. Science is one of its expressions, and a system of universal human rights—the famous 'Declaration of the Rights of Man' from the era of the French Revolution—is another."

"I don't want to quibble, Hera, but may I point out that

women weren't included among the beneficiaries of the Declaration?"

"I concede the point, and I could also seek to turn the charge against your faith, but I won't. The work of reason is sequential. Science didn't spring fully formed from the minds of its great eighteenth-century originators, and neither did modern democracy. What was set in motion then was a self-developing organism that unfolds and expands over time."

He smiled. "How does the saying go? I was just trying to yank your chain ever so gently."

"I should have known. But will you now throw in the towel? My science doesn't arrive naked and unadorned in the world. It, too, appears as part of a larger package of goods. The package also contains an entirely new kind of governance structure for society, one that includes institutions where reasoned debates about what is fair and appropriate can occur. Those debates provide, among other things, a legal and ethical framework for human actions. Its chief difference from what came before is that it isn't frozen in time, like your Ten Commandments, but rather evolves continuously to meet novel challenges. Among them are issues arising out of the bestowal of new powers on us by the sciences."

"As I have learned to expect, you almost always come up with a statement that is as eloquent as it is persuasive. But permit me to probe a bit deeper. Here is what still bothers me. Science itself and its practitioners seem to get off scot-free in your account. The mandate they labor under is, apparently, to generate a steady and presumably unending stream of new knowledge and new powers. At that point, some entirely separate and distinct social mechanism is supposed to kick in, and those 'debates,' as you style them, occur in some other forum. Have I represented your conception adequately?"

"Yes, you have. A theory of society grew up to explain it. One can refer to all of the various aspects of this social order as 'subsystems'; among them would be the economy, the private sphere—for example, the family—politics, law, religion, and science, too. A modern society is inherently disaggregated. That's one very important way in which it differs from what are called traditional societies."

"I see. Let's continue to concentrate on science for a moment, shall we? Do I correctly infer, from your story about how this modern social order operates, that scientists would bear no special responsibility in the matter? Thus they have had no special duty to ensure that those 'debates' you mentioned actually take place? Or that the content and timing of the debates would be adequate for the purpose they're supposed to serve?"

"What you say does follow from the theory. All individuals have a multiplicity of roles and responsibilities in society. All scientists are also citizens of the polity, of course, and it is in this capacity that scientists would have participated in those sessions, along with whomever else wished to contribute to them—on a purely voluntary basis, of course."

"They couldn't have been compelled to take a prominent part?"

"No, with a few minor exceptions; for example, if there were court proceedings, or if they had been required to do so as part of an application for research funding, but such cases were rare."

"I must say I find the situation a bit odd. Scientists are the original creators of this hugely expanded set of capacities for action in the world. How could they not bear a very particular responsibility for overseeing the uses and outcomes stemming from their discoveries?"

"The great majority of them took their cue from Francis Bacon, an influential thinker of the seventeenth century. He convinced them that scientific knowledge is innocent."

"And what about power?"

"In so far as knowledge gives rise to power in the new sciences—the powers inherent in nature and now revealed—the implication in his thinking is that power, as such, partakes of the innocent character of that knowledge. Only when it's actually used to some end can a judgment of good or ill be applied."

"Let's choose an example. You've talked to me many times about neurosciences and the manipulation of brain function, since you and your sisters have personal experience in this regard. So pick an example from that field."

"That's easily done for me! All right, I choose memory, in part because absolutely everyone knows what it is, and many have watched aging parents and relatives be challenged by memory loss. To be sure, in anything to do with potential medical applications, there is at once an intensely practical interest as well as a purely intellectual challenge. But we can begin with the latter aspect, where the question posed is exactly the same, no matter what field of scientific inquiry we mention: How does the process work?"

"Since memory is a part of our minds and brains, the answer must lie therein, I assume."

"Indeed, that is so. Eventually science knew that brain functions occur in specific types of neurological cells and also, for the most part, in specific regions of the brain. The process involves both chemical reactions and electrical impulses, and it can be regulated by gene expression. Various malfunctions can occur as a result of organic damage from accidents or disease, from inherited genetic disorders, from psychoactive chemical substances, and other causes."

"Now we know how memory works, if not completely, then to some extent. Now what?"

"Now we're ready to craft some therapeutic remedies for people who are suffering from a disease or condition that's preventing them from functioning normally. Take the case of traumatic memory. The brain is an emotionally charged organ. It has powerful responses to psychological traumas, such as the experience of violent rape. The victim doesn't just suffer once from the impact of the original crime. In many cases the victim's brain spontaneously revisits the horrific memory of it time and again. It's a truly awful form of never-ending secondary punishment."

"Can such a person be helped?"

"The answer is yes. Researchers found a way to use a chemical to block the recurrence of those specific memories without interfering with normal memory function. It was miraculous. And then there were other applications. For example, almost immediately, the therapy was proposed as a prophylactic treatment for soldiers who were about to go into combat. In situations where civilian casualties were inevitable, the military thought it would be useful to inoculate their warriors in advance of the events. In other words, they figured out how to selectively block certain types of recollections from being readily 'consolidated' in the brain's circuitry where long-term memories are stored. Critics of the procedure accused the perpetrators of creating zombies, human killing machines who would be utterly unaffected by the horrors they created or witnessed."

"You're saying that this type of application was also successful?"

"Who really knows? All we are reliably informed about is that the treatment was carried out in many different countries. What's important in this context is how quickly a whole host of problematic applications were spun off of the original benevolent intention."

"Certainly any specialist in the science of memory could easily envision not only the likely benevolent objectives, but at least some of the malevolent ones as well."

"I think that's a reasonable assumption to make."

"So here's the scenario we've just outlined: Science discovers the process of memory formation in the human brain and …."

"More accurately, in the animal brain. Ours is just a highly evolved version of the basic mammalian model."

"Fine, in the mammalian brain, then. Some potential uses to which this knowledge can be put are immediately apparent."

"Some of them may, if successfully commercialized, make the discoverer and his backers very wealthy, too."

"True enough. Not necessarily a completely ignoble motivation, I suppose. Whether or not it has this result doesn't affect the point I was going to make."

"Sorry!"

"Forget about it. The point is, being able to reconfigure memory is a potent tool, to say the least, because technologies that can manipulate our minds are pushing into very sensitive terrain. Based on what you told me earlier, the scientific discoverer would have fully discharged his formal obligations to society if he or she shoved a copy of the relevant publications across the table to the rest of us and whispered, 'Now, you'll be careful with this little device, won't you?'"

She laughed merrily. "What a sharp wit you have hidden beneath your somewhat dour exterior, Jacob! Actually, most don't bother to issue the warning."

"Whatever. The bottom line is, the discoverer's responsibility seems to end at the point when the publication appears."

"That's quite correct, actually, because scientists see their primary duty as remaining true to the method they're all supposed to share and uphold. When they have results to report, they're obliged to do so in full public view as a way to demonstrate their fealty to the approved method. Whatever happens afterward is somebody else's worry."

"I must say I find such an attitude to be most unsettling. At the very least, it seems to me, the discoverer ought to assume another type of responsibility entirely. This might entail preparing something like a prospectus covering the potential uses, together with a discussion about the nature of any problematic issues that might arise from them."

"That's not the way the system worked, until recently anyway. In large part the reasons had to do with the enormous competitive pressures that arose once science went global. And with the fact that the potential financial rewards to the inventors multiplied by orders of magnitude. National prestige, too, played a role, especially with respect to newer players in the game, such as the Asian countries."

"Even though you're such a partisan of the scientific method, I'd be amazed if you told me that you approve of the way the system is working. Do you?"

"*Mea culpa,* you smoked me out. I always feel caught in a bit of a bind when I've set myself up to represent science in opposition to faith. I am, in truth, an unabashed devotee of the scientific method. But yes, I concede that there is a serious problem here."

"Describe the nature of the problem to me."

"It's a function of the historical trajectory I mentioned earlier. Modern society separates the sphere of scientific activity on the one hand, with the oversight of the potential impacts of the new powers resulting from that activity on the other. The net result is pretty much as you characterized it a moment ago. The scientist uncovers how some natural process works and says to his society, 'Well, here are the results, isn't this just marvelous? Look at all the things you can do now that you couldn't do before. Of course, you must remember that what you decide to do is your responsibility, not mine.'"

"I'm pleased that you've adopted my terminology. Now let me ask: Recalling to mind your own earlier metaphor, I wonder whether what you've just described isn't a kind of 'black hole' also? What I mean is this. A zone of astonishingly potent forces governing the behavior of matter and energy suddenly appears in our midst. In the absence of modern science, we would never have known these forces exist, how they work, or how we can turn them to our own advantage. Now this new knowledge sits there, available to everyone, and anyone can dip into it, anytime into the foreseeable future, for any purpose whatsoever. Your scientist, having unveiled it and made a bequest of it to us, walks away and commences a new investigation. I realize we are speaking in metaphors, but that sounds like a metaphorical black hole to me."

She sat there gathering her thoughts for some moments. "I concede that a point in time was reached when a price began to be exacted for science's refusal to acknowledge its own share of responsibility for overseeing the vast expansion of human powers. Thereafter, the price to be paid for science's failure to impose limitations on its applications has been multiplying exponentially."

"You mention a point in time. When was it?"

"During the First World War."

"Hera, that was a century and a half ago!"

"True enough. I hasten that some brave attempts were made to alter course, but they failed."

"I would love to hear more, but other duties call me. Perhaps we can pick up this thread again sometime later."

"Welcome home, sister. Welcome to our magic mountain. The band of sisters is made whole again," Hera murmured, tears streaming down her cheeks as she held Io tightly to her. Io and I had just disembarked at our private airstrip located in Jackass Flats near the eastern slopes of Yucca Mountain. When word had reached Hera and her sisters two weeks ago that Io was finally ready to leave the family estate in Bali, after nearly a decade of convalescence, and to rejoin her siblings at their hideaway in the deserts of the southwestern United States, I was despatched at once to serve as escort. Of course, I needed no order or incentive to prompt me to fetch my own mother back to her tribe. During the long years of her exile, our hopes swelled and ebbed like the daily tides as we waited to learn whether the long-term care we had arranged at her grandparents' secluded sanctuary would subdue the severe schizophrenia afflicting her.

No journey halfway around the planet's surface is either simple or relaxing these days, and because there's almost no commercial air traffic in Asia anymore, only our contractual partnership with the US Air Force made the trip possible at all. Ever since the Indonesian government collapsed for good about a decade ago, the island group in the Malay Archipelago known as the Lesser Sunda Islands, stretching from Bali to Timor, has been an Australian protectorate under United Nations authorization. Having flown to Sydney and then on to Darwin, I waited a few days until I was picked up by a group of Australian naval vessels heading for the Bali area on routine patrol. Once we had arrived in the narrow strait separating the islands of Java and Bali, an onboard helicopter flew me into the mountainous terrain of western Bali, past Mount Agung, landing at the Sujana family's estate, where Io was waiting for me. We then retraced my steps and arrived at Vandenberg Air Force Base on the Central California coast, where we were met by our executive jet and taken immediately to Yucca Settlement.

As Hera released her grip, Io took a step backward and, turning in my direction, said, "My dear sister, I'm so sorry. Forgive my rudeness. I haven't introduced you properly to my handsome son. This is Marco, my brave rescuer."

I glanced at Hera, whose face betrayed not the slightest sign of alarm as she replied, "Oh yes, I've met Marco. It has just slipped your mind, sister." Was it the anti-psychotic therapies, I wondered, or the illness itself, or both, that had left these gaping holes in her memories? Or was her brain protecting her—and itself—by blocking all recollection of events starting with the onset of those nightmarish years when paranoid schizophrenia ran unchecked through her neurological circuits?

Hera now turned and motioned the rest of the sisters to come forward. They approached, grouped into subunits according to the work they did, which at first struck me as odd, until I asked myself: Why not? This is how they spend most of their days and nights.

"And here's the rest of our little band, dear Io." First came the scientists and directors of the microbiological labs, heirs to their famous father's training: Gaia, Pandora, Persephone, and Rhea. Then the administrators and market traders, in whose hands the Sujana Foundation's huge financial holdings are entrusted: Athena, Artemis, and Hecate. Finally, the wizards at mathematics, cryptography, and computer programming: Ariadne, Moira, and Themis. After each had embraced her in turn, Gaia said, "Once you've acclimatized yourself, Io, we'll begin to make the introductions to our precious tribe of youngsters, now almost eighteen years old."

Later that evening, we sat on the veranda extending off Hera's study, one of her suite of rooms dug into a low hillside at ground level, below the crest of Yucca Mountain, at the edge of Crater Flat. The modular sections of flooring, walls, and doorways salvaged from her Japanese-style house in the Caribbean now grace her desert home. Around the veranda's edge, a small stream, run by pumps only at night and coursing over a bed of stones and broken here and there by low waterfalls, meanders through the beds of desert flowers and grasses in the surrounding garden. She had often remarked on the strange sense of peacefulness induced in her by the sound of trickling water.

Beyond the garden stretches her collection of cacti and desert shrubs; further away are visible our massive plantations of fruit and nut trees, extending in all directions across Crater Flat. We could hear the muffled chatter of the apes bedded down in their nearby enclosures. Above us a brilliant moon cast innumerable shadows across the ground and illuminated the darting, scurrying figures of its indigenous nocturnal mammals. Away in the distance the nightly predators' concert commenced, the voices of foxes, wolves, and coyotes alternating in choral recitative.

Hera and Io reclined side-by-side in chaise lounges on the veranda beneath the moonlit sky. "Go back to the beginning," Io insisted. "I remember a village where we grew up, and it must have been in or near Bali because the servants on the Sujana family estate looked just like the ones I remember from our childhood. But where have you been living since we were little girls, Hera? Have you been here in the desert all this time?"

Her sister was holding one of her hands as she spoke. "No, we were all together until about fifteen years ago. We did grow up in an Indonesian village set on a great mountainside, a village named Tembagapura, that much is correct, although not in Bali, but nearby, in Papua Province. We never knew our mother, who died before we were born, but our father was with us there, and also later, during the time we spent living in the Turks and Caicos Islands, in the Caribbean, where he passed away some years ago."

Io looked at me, smiling, and reached out her other hand, which I took in mine. "And so my precious son was born in the Caribbean?"

"Marco was still a baby when we moved to Providenciales. There was a stop in between, Io, directly north of where we are now, in Canada, in southern Alberta. We lived on a ranch near a small town called Longview. You gave birth to Marco in the nearby city of Calgary."

"It's quite awful to have these blank spots; I've no recollection of being in a place called Alberta—or of islands in the sea, either. But no matter. Now that we're back together you can tell me all your stories, and then my memories will be triggered."

I sincerely hope not, I thought to myself. She's better off without them.

They were silent for some moments until Io spoke again. "Gaia said something odd earlier, about a big group of teenagers who were staying here. Are you running something like a summer camp, Hera?"

She laughed merrily. "Oh yes, often my sisters and I do seem to be just like camp counselors, sister! I must remember to tell the others what you just remarked—I'm sure they'll find it as funny as I did!"

The next morning Hera invited Io and me to the study in her suite. A large topographical map in shaded raised relief, displaying Yucca Mountain and its environs, was laid out on a table. Nearby was an entire wall covered with aerial photographs and satellite images that revealed the inventory of our built structures, crops, vines, greenhouses, and tree plantations in startling detail. "Come and see these displays, sister. In the future this will be referred to as our Mother Settlement, the place where our kind took root and began to flourish."

Also hanging on an adjacent wall was a flat map of the region within a 100-mile radius of our encampment. "We're living in the Mojave Desert—a very, very arid desert, as it happens, with average rainfalls of well under two inches a year. But don't despair! By a stroke of luck, you've arrived in springtime, when the gorgeous wildflowers bloom. We have hundreds of different spring annuals, many of them unique to the southwestern deserts. The shrubs and cacti bloom at the same time."

"But how do you manage to grow all these plants without water?" Io asked.

"Ah yes, water, the essential of life. Already by the end of the nineteenth century both the cities of San Francisco and Los Angeles required imported water, delivered from sources in the Sierra Nevada mountains by the first of two long watercourses, the Hetch Hetchy and Los Angeles Aqueducts. By the middle of the twentieth century, the population and the agricultural industry of the entire region of the southwest—southern California, here in the Las Vegas area, and Arizona—began to explode, whereupon the Colorado River was called upon to do its duty. This entire region is naturally dry, at least in the current geological time frame, and its indigenous water stocks would support only a fraction of those numbers.

"So four new, massive aqueducts were built—three sucking off the Colorado River, including the Colorado Aqueduct and the All-American Canal, both flowing east into California, and the Central Arizona, going west to Phoenix and Tucson; and one, the California Aqueduct, running north–south and sourced from the Sacramento River east of San Francisco. Around the turn of the twenty-first century, something like three trillion gallons of water per year were flowing into the arid southwest, two-thirds of it contributed by the Colorado River. For a few decades, everyone was happy: huge vegetable crops were disgorged from irrigated farms in the San Joaquin and Imperial Valleys; water cascaded freely in the outdoor ornamental fountains at the Las Vegas casinos; green grass sprouted at hundreds of golf courses and on tens of millions of lawns; countless artificial ponds and streams graced the gated communities; all constructed in full view of the surrounding bone-dry deserts."

Hera deferred to me when I indicated a desire to continue the story.

"Then, beginning early in the second quarter of this century, as predicted by climate science, the entire region's rainfall failed almost completely. In response to persistent drought at its headwaters in Colorado, Wyoming, and Utah, the Colorado River's flow plummeted and never recovered. Wild imprecations were hurled at nature's lack of appreciation for all that humans had done to turn a roasting wasteland into a calm oasis of graceful and opulent living. But quickly the hatred was redirected toward one's human competitors, and at first what counted most was how far upstream you happened to live.

"In an act of desperation, the Nevada government shut off the flow of water in the Colorado River at Lake Mead, just east of Las Vegas. That left California's Parker Dam on Lake Havasu—which since 1960 had been pumping one billion gallons per day through the Colorado Aqueduct into southern California cities, and half-again that much eastward to Arizona—high and dry, so to speak. Those citizens furthest downstream, the former beneficiaries of the All-American Canal in the Imperial Valley, were the first to dry up, ignoring the Mexicans, of course, which was standard practice in any case.

"Lawsuits were soon replaced by private vigilante justice, a pattern seen first in the case of the Los Angeles Aqueduct, perhaps because the historic grievances of people in the Owens Valley—where the northern terminus of this aqueduct initially was located—were so much older. As social chaos spread, both private diversions and, much worse, simple acts of sabotage were impossible to control along the hundreds of miles of its length, much of which was sparsely settled.

"In the case of those channels fed by the Colorado River, the authorities' initial response was to beef up military security along the route, but alas, when the river gave out, there was nothing to guard. The exodus began, soon swelling into a great back-migration to the eastern United States, where those who had envied their suntanned lifestyles from afar now found it irresistible to mock and humiliate the returnees. Far less time was required for the region to empty out than to populate it in the first place."

"Some evening soon," Hera interjected, "we'll watch an old movie together, Io. It's called *Chinatown*. It depicts some of the ugly episodes that took place when the Los Angeles Aqueduct was being built. Although I have to say, what happened when the aqueducts started to go dry was even uglier."

Io had listened to these disquisitions in silence, with a bemused expression on her face and occasional glances at the maps and pictures. Now she looked at Hera. "What a fascinating story, sister. What puzzles me is why we should want to arrive here when everyone else is taking off. So I ask again: Have we figured out a way to exist without water?"

She smiled. "Look at this map, Io, which shows the area directly west of our location at Yucca Mountain. If you trace a straight line from Beatty, the little town nearby that we use as our administrative center, and go fewer than eighty miles, right through the northern end of Death Valley, you wind up at Owens Valley, the original source of the water diverted into the Los Angeles Aqueduct. The local ecosystem is a lot healthier since the old aqueduct project collapsed, but there was a significant loss of income in the area, too. That's when we stepped in and offered to take a much smaller quantity for our needs on a long-term contract. When the folks in Owens Valley agreed to the deal, we had an underground pipeline constructed across Death Valley and we placed our own permanent security contingent around the only vulnerable point, which is the intake facility on the Owens River. At the other end is our water treatment plant in Beatty."

"The other photos show what we can do with a good water supply in the desert," I added. "Look here." I was pointing to the pictures showing our fields of pulse, cereal, and oilseed crops arrayed in Crater Flat. "We get most of our nourishment from the cereals and pulses, which also connect us nicely to our human past in the Americas, since the indigenous peoples of the New World first cultivated them eight thousand years ago. Through buried drip lines we irrigate directly at the plant roots, so as to avoid the otherwise unavoidable evaporation losses from irrigation ditches, and we do the same with our tree plantations. Our many types of nut trees provide another good protein source. We got lots of good advice from the team of Israeli agronomists who spent a year with us during our setup period.

"The fruits of citrus and avocado trees are consumed fresh, and we dry most of the soft fruits for winter eating. We have date palms and the native desert palm, as well as cottonwoods for shade. Our vegetables and herbs come from those long glass structures there, which are hydroponic greenhouses, although we have some outdoor gardens as well, including our precious vineyards and olive groves! We have fish farms and we keep herds of sheep, goats, and cows for our cheese supply, and chickens for eggs; but as you may remember, sister, we don't eat the meat of land animals."

"Where do you keep the farm laborers?" Io inquired.

"You're looking at two of them," I replied. "Everyone in our Settlement has a daily quota of activities in food production, which is carried out under the supervision of our resident Hutterite communities. Our operations also supply the group of technicians and medical staff, and their families, who live either here on-site or at our administrative center in Beatty, all of whom also take their turns in the fields, groves, barns, and greenhouses.

"We use machinery, automation, and computer-controlled processes wherever we can, of course, so we're not talking about traditional backbreaking labor here. But many of us actually prefer getting our hands dirty, as the saying goes, enjoying the sheer tactile pleasure of handling the soil, plants, and animals. For our youngsters, the farm routines provide a welcome break from their grueling study schedules."

"We actually have a substantial surplus at the moment," Hera added, "since we've been gearing up for a major population expansion when our teens start having children a few years hence. In the meantime, we can satisfy most of the food needs of the several thousand people, scientists, lab technicians, and graduate students, including their families, who live and work at Sujana University in what's left of Las Vegas. We administer that area as caretakers, on behalf of the US government, jointly with the military command at Nellis Air Force Base, which is our contractual partner in this region, in anticipation of the day when the water levels and rightful property owners return, when the roulette machines reawaken to gorge themselves again on nickels and quarters through the long nights when no one sleeps—a series of improbable events that, we're quite convinced, isn't going to transpire anytime soon."

I broke in. "Let's go for lunch. Then Io can relax while you and I have our meeting with the security committee. We can look at some more of the exhibits later today."

When we resumed later that day I pulled out the pile of old construction drawings for the site. The top page contained the master list of facilities, which I passed to Io:

A. Internal (Mountain and hillside façades, ground-level, Crater Flat perimeter):
   1. Stacked modular residences, main compound: 1,000+ units
   2. Residences, main compound (sealed; future growth): 3,000 units
   3. Residences, Hutterite compound: family units
   4. Residences, on-site technical/professional staff: 500 units
   5. Communal facilities (dining, recreation, education)
   6. Barns (pens, stalls, feed/hay storage)
   7. Ape range and caves
   8. Internal plumbing, wiring, ventilation, light-pipe conduits
   9. Elevators, access ramps, and interior/underground connecting tunnels

B. Underground Tunnels (Vault Area, Yucca Mountain):
   1. Small-scale nuclear power plant

2. Mainframe computer and process control panels
3. Hospital and clinic
4. Air quality monitoring and filtration/circulation station
5. Backup water testing and treatment plant
6. Recycling, waste treatment and incineration plant
7. Labs, machine shops, repair and maintenance bays
8. Refrigeration/humidifier – dehumidifier plant
9. Flywheels and storage battery array

C. External:

1. Landing-strip, hangar, and helicopter pads
2. Exercise and playing fields (lighted)
3. Interior roadways and connecting underground tunnels, with vehicle bays
4. Weapons emplacements, perimeter fence, cameras, sensors
5. Agricultural fields and tree plantations
6. Solar panel and windmill arrays
7. Hydroponic greenhouse arrays
8. Buried water piping
9. Lighting, antennae/radio towers, radar/communications dishes

Io put her finger on the list and smiled. "Are you serious? A nuclear power plant? Didn't you tell me last night that our facilities are grouped in close proximity to a huge quantity of high-level nuclear waste buried inside the mountain we're now standing under?"

"I'll get to the nuclear waste in a moment," Hera said. "The nuclear power plant serves as backup, in case our solar and wind power arrays were to be destroyed in a severe storm. The reactor is a compact unit cooled by helium. Very ingenious. But under normal conditions we don't need it, so it's kept on standby status. The desert is windy as well as dry, and between the solar panels and our wind turbines we have a surplus of electrical energy on many days. The surplus daytime power drives a set of enormous flywheels, which store the energy, and as they run down all night, they recharge our storage batteries. In fact, we have so much electricity that would otherwise go to waste that we often work the fields and play sports under floodlights at night, when it's a lot cooler."

Io interrupted. "So what about the stuff that's underneath the mountain? Isn't what's stored there said to be the most dangerous material on the planet? And you've got us all living right next door to it?"

"I know it sounds a bit ironic," Hera replied, "but very few remaining major facilities still standing in the United States have been engineered to as high a level of safety as our repository here. We even draw practical benefits from it. It gets pretty cold in this desert during the nights and throughout the winter, and it so happens that the waste generates a lot of heat that's vented off the storage canisters, which we redirect to keep us warm and provide hot water. The heat will last for something like three hundred years! There are also plenty of empty underground vaults inside the mountain in which we've stored our most critical facilities, such as our central computer, the nuclear reactor, our hospital, and our secret microbiological labs.

"It's the final move for us, Io. Our long and perilous Odyssey is over. Once we've finished settling around Yucca Mountain, we'll be staying, no matter what happens in the future. There are only two options anyway. The Amargosa River Valley will become for our species what Africa's Rift Valley now means for humans—the point in time and space where its great adventure began. Or it will become the mass graveyard of the Second Generation. There's no third option in these matters."

Lucetta, *luce,* light, lustrous in the divine rays of the morning and the evening sun, a rich lustre enveloping me by day, a luminosity lighting my way beneath the starry canopy at nightfall. How could it have come to pass that you selected me? Already at age seventeen you were an exceptionally bright beacon on the many-starred canvas of the Yucca tribe's Second Generation, destined for glory among that company with whom I must foreswear all intimacy, or so I thought. And then you were on my doorstep one day, and it was plain that you would have me, and I was doomed.

§ § §

"Those little shits!" Lucetta yelled, to no one in particular, as she arrived at my comfortable trailer in Beatty one day and yanked open the door.

"Hello to you, too," I replied. "Which of your esteemed associates are you referring to, may I ask?" I received a very nice kiss by way of apology for her manner of entrance.

"You know Ming, right?" I nodded.

"He's a good friend, I really like him. Well, this morning he was working next to me, mucking out the horse stalls, and told me this story:

'They've got you in their sights, Lucetta; I assume you know that?'

'Who?'

'The little group of boys—perhaps I should refer to them more respectfully as young gentlemen—who call themselves the Silverbacks.'

'You've got to be kidding. Silverbacks?'

'I note that you find the matter to be amusing. Well, maybe it's funny, maybe not at all. I thought it was pretty comical at first, when they tried to recruit me. Perhaps it was the list of grievances they had drawn up. There was some stuff about how it was time for the old ladies to retire from the scene, meaning Hera and her sisters, of course. But mostly it was about being outnumbered by the females of their own age in the Assembly, and having to listen to 'tedious harangues'—their words, not mine—by some of the women, especially you and one or two others. I believe the appellation "harridans" might have been used.'

'Assholes! When did this attempted recruitment take place, Ming? Who approached you? And what did you say?'

'It was last week, and the emissaries were Rainer and that constant sidekick of his, Kenji. I didn't need to reply because apparently I didn't succeed in suppressing the bemused look on my face. I was told that they didn't need me, that I should run off to my sewing class, and that I should just stay out of their way. When I asked what that meant, the answer was, "You'll find out when the time comes."'"

I looked at Lucetta, who had been marching up and down in the hallway of my modest dwelling as she relayed the story. "I did pick up rumors recently about some such group, but frankly I didn't pay much attention. It sounded like just a few of the lads blowing off steam and metabolizing some excess testosterone. Was I wrong?"

"Wait until I relate my own brief conversation with the emissaries," she added. "I strolled up, unable to suppress a grin, I confess, and introduced myself as president of the Harridans Club. I said that they could join my club if I could join theirs. Rainer's little puppy Kenji started laughing, until a glance from his master shut him up fast. Rainer gave me a condescending look and said something like this:

'I think it's best for all concerned if we keep the membership lists of our two little groups separate, Lucetta. Besides, certain biological obstacles make it awkward to try to carry out what you've proposed.'

'Oh, you must mean that women appear to have a harder time than some men do in figuring out how to allow their inner gorilla to emerge.'

'Very funny, I'm sure. Actually, I was referring to the obstacles preventing our joining your club.'

'There I do have to concede your point. It seems indeed that the designation harridan—as well as shrew, its pithier synonym—is reserved for women alone. Don't you find it just a little bit interesting, Rainer, that there's no equivalent expression for the male?'

'A fascinating linguistic conundrum, I agree, but we don't have time to improve your command of the English language just now. Nevertheless, we should both make an effort to stay on good terms with each other. I'm hoping to have full support from you and your followers for the motion we'll be introducing in the Assembly soon.'

"I waited, because I knew he couldn't resist telling me more even if I didn't take the bait. Kenji blurted it out before Rainer could stop him:

'That would be the motion providing for votes to be weighted, using a formula that will give the male and female caucuses as a whole an equivalent number of votes.'

"As I stood there dumfounded, Rainer took his puppy by the arm and strode off."

"Have you told Hera about the contents of this little chat?" I asked.

"No, not yet, I haven't had a chance. I did talk to some of my friends; like you, they didn't seem to take it seriously. That may be a mistake."

"The Assembly is three-quarters female. Surely you don't think that enough of the women will go along with the idea to pass the motion. So that will be the end of the matter."

"Will it? I wonder. Is it inconceivable to think that they might have a backup plan?"

At that moment Ciso knocked on the door and I invited him in. "Hi, Lucetta." And then to us both: "I'm done with my chores for the day and I'm thinking of taking Magnus for a walk in the hills. Care to join me?"

We all left together.

§ § §

She was apoplectic. "How *dare* you? Do you imagine that you are Yucca's harem-keeper? That you're entitled to take your pick of whomever you fancy among our young women? You're twice their age. Have you no shame?"

We had met later that evening in Hera's apartment for what I had thought was going to be a routine review of administrative business, and I was completely unprepared for her onslaught. What an utter fool, I reflected later. It had never once occurred to me that my obviously innocent dalliance could take on the proportions of an affair of state. Truth be told, I had been proceeding trance-like through my daily routines ever since the day that Lucetta had first walked me to my bed. Whatever qualities of judgment I normally possessed had been taken off-line at that moment, hijacked and hidden away so expertly and surreptitiously that I had no idea they were even missing. The need for her filled every crevice of the spaces where my now-superfluous neural circuitry had once resided. There was no way I was going to win this argument, and I was not prepared to even consider losing it. So I turned on my heel without a word and fled the scene as Hera screamed after me, "I forbid it! Do you hear me? I *forbid* it."

Narciso apparently heard me roar into the Beatty compound and slam the door on my trailer, set adjacent to his, because a few minutes later there was a knock at my door. He took one look at me and asked, "What's wrong?"

"None of your business. I'd rather be alone if it's all the same to you."

A short time later I walked over to his place, carrying a bottle of tequila as a peace offering, and told him what had happened. When I finished, he said, "I don't usually like to get involved in these delicate matters, but, for what it's worth, I'll say my piece if you like."

"Go ahead."

"Let it rest for a bit. I've gotten to know your lady friend, and I've watched you two together. There's no way that even the formidable Hera is going to get away with telling Lucetta whom she can fall in love with. Let's go down to the Sourdough and see if we can take your mind off your troubles."

The saloon was jam-packed, but we managed to find two empty stools at the bar. Usually I know when to stop drinking, but that day was different, and not even Narciso's gentle admonitions mattered. Finally, he signaled the bartender and stood up. As he hauled me off my stool, I lost my balance and stumbled backward, colliding with a young lout just passing behind me. I was about to mumble an apology when he shoved me away roughly, saying, "Watch where you're going, dickhead." In milliseconds my rage soared and I hauled back my arm. Narciso was quicker. Only later, following my recovery, did I discover what he had done. Stepping between us, and sweeping his arms across in opposite directions, he had struck each of us on the bridge of the nose, just below the forehead, with the edge of his unclenched hand, whereupon we both dropped to the floor unconscious.

I awoke back in my trailer to Lucetta's gentle ministrations. As she removed the ice-pack from my forehead she made some remark about hormones, but fortunately I was still dazed, and only later did I realize that I had missed the full richness of her sarcastic tone. When I had pretty much recovered, she glanced at me.

"I was looking for you and ran into Narciso outside the Sourdough, just after he had deposited your limp form in the cab of his truck. I found out from him what happened and roared straight back to Hera's suite. She didn't answer the door until I pounded on it, and when she finally appeared I could tell she'd been crying. That didn't stop me. We had a furious row. As I left I dared her to summon her police force and throw me into the Beatty jail because that was the only way she could stop me from seeing you."

I started to mumble a reply, but she put her finger across my lips. "Of course, she may have scared you off, and if so, there's not much I can do about that."

"Why don't you play the Rolling Stones' 'Wild Horses' for us?" I whispered weakly.

"I know the lyrics," she replied, smiling sweetly.

Part Two

Jackrabbit Spring

Einstein:
But all the same we cannot escape our responsibilities. We are providing humanity with colossal sources of power. That gives us the right to impose conditions. If we are physicists, then we must become power politicians. We must decide in whose favor we shall apply our knowledge.... Our political system too must eat out of our hand.

Friedrich Dürrenmatt
The Physicists (1962)

In late May, Narciso and I had returned from a quick trip to Europe, having hitched a ride out of Nellis Air Force Base on a military transport, going non-stop to Rome, thence courtesy of the Italian Air Force to Sicily, where a fast boat was waiting at Siracusa to take us to Malta. Over the preceding few years, a bevy of scientists and their families, originating in various failed countries around the rim of the Mediterranean Sea, had washed ashore at Valletta—with a little help from our agents in the area. Narciso and I collected the latest batch, and with the group in tow, we retraced our steps to Las Vegas.

About a week later, I'd been reviewing our next assignment with Gaia when she glanced at her watch and said, "I've got an appointment with one of the fellows you rescued, Abdullah al-Dini. He appears to have some concerns about how things are run here. I want you to sit in on the meeting. I'm also expecting Athena to drop in soon, and when we're done here we can all have some lunch together."

The new arrival walked through the open doorway into Gaia's office at that moment, and there followed some brief exchanges of pleasantries.

"Perhaps I should begin by reviewing my professional career," he remarked.

"You don't have to present your credentials to me because I'm thoroughly familiar with them, Dr. al-Dini. We went out of our way to recruit you precisely because of your stellar reputation in molecular genetics. But the message you sent yesterday requesting this meeting leads me to believe that you think we've brought you here to join the faculty under false pretenses. I'm extremely sorry to hear you say this. I had thought I disclosed everything about how our Las Vegas operations are run in the contract you signed. Didn't you read it?"

"The situation at Al-Fateh University in Tripoli was very bad just before my family and I went into hiding and escaped by boat to Malta—thanks to the assistance you provided, Madam Gaia. Spying on our laboratories by the religious authorities had increased over the years, and more recently we were faced with demands to pursue certain lines of research that we considered inappropriate. Even after arriving in Malta we were preoccupied with our safety. I did sign the contract there, but I confess I didn't examine it until after we were settled here.

"Before we turn to the issues that occasioned my message, may I ask first about the rather curious name of your establishment, Madam Gaia?"

She smiled. "Please call me 'Gaia,' Professor. 'Solomon's House' is indeed our official designation, although the staff here immediately started referring to it simply as 'The House.' Our leader, Hera, is a devoted acolyte of the English philosopher Francis Bacon who, early in the seventeenth century, wrote a short book about an imaginary scientific establishment, which he baptized 'Solomon's House, or the College of the Six Days' Works.' You may recognize the Biblical reference in the title, which rather amused the rest of us, but we like to indulge Hera a bit, so we agreed.

"Anyway, we were pleased to be of help to you and your family. Just tell me what aspects of the contract concern you."

"My concerns have to do with the standard procedures for publication and dissemination of scientific research articles. Since you are not a scientist, you may be unfamiliar with them."

I glanced at our visitor and waited for Gaia's riposte.

"Dr. al-Dini, I may not have academic degrees or research publications, and I have never been a university professor, but please do not assume I'm ignorant of those procedures. On the other hand, perhaps I shouldn't blame you for making such an assumption. There is no way you could have known that I spent fifteen years in intensive, private training and collaboration with my father. You also could not have been cognizant of my tutor's identity and my bloodline: I am a daughter of Franklin Stone and Ina Sujana."

"Forgive me, please! Of course I recognize those names. I did not intend to insult you. Please accept my apologies."

"There's no need to apologize. We don't broadcast the news of our lineage. Let's just return to the issue at hand, shall we?"

"I want to make my position crystal-clear. My family and I are very grateful to be here, safely away from the turmoil in our native land. I have no objection to the other rules applicable to everyone here, and the reasons for them, for example, why you do not offer salaries."

"We were pretty sure that scientists could persuade their families to get by without shopping malls and designer outfits, Professor, if the choice was between having either those amenities or adequate research funding. For the same reason, everyone must work, as you are aware, starting at about age ten, in addition to the time allocated to all for schooling or training, of course. We earn a very good income from our commercial operations in the manufacture of medical and scientific supplies, and every bit of it goes to subsidize the R & D activities at Solomon's House."

"As for the living arrangements, Gaia, we do appreciate having a private apartment in the hotel, and we don't mind taking our meals in a communal setting, or receiving articles of clothing and other personal necessities at the dispensary. But I admit I was shocked when I first met the scientific staff here, including the faculty, and found many of them outfitted in dishdash robes! My mind suddenly flashed back to my origins and for a moment I was quite disoriented."

"We believe we should dress appropriately for a desert climate, Dr. al-Dini. So, yes, we offer a nice selection of robes, in a variety of styles and fabrics for both men and women, including Arab robes and, for those who prefer, tunics and togas in the style of ancient Greece and Rome. But you won't find any hijabs in our shops, I'm afraid; just sturdy wide-brimmed hats to protect against the fierce sun."

"What bothers me is just one point. You have a huge professional staff in molecular biology here, hundreds of the best scientists in the world, as well as advanced students and skilled technicians. But we aren't allowed to publish or disseminate our results outside the College. Surely this is a violation of the spirit of scientific inquiry."

"True, I concede your point, but in every other respect the rules are the same. You are obliged to conduct research and to publish the results in the electronic journals we circulate at the College. All decisions, from hiring of personnel to the allocation of research budgets to oversight of research ethics, are made by committees of scientists, without outside interference. The sole exception is the privilege that is reserved to our parent body, the Sujana Foundation, to designate priority areas of research deemed vital to its interests, such as virology. As for publication in our journals, peer review rules the day here as much as elsewhere and there are no restrictions whatsoever on internal communications."

"All that is fine. It does not, however, explain or justify the restriction you impose on external dissemination of our results and on communications with scientists working elsewhere."

"Not everyone's field is affected, Professor. With a few exceptions, the restrictions apply mainly to biology and genetics, as well as a few areas in chemistry and physics."

"And the justification is—what?"

"How shall I put this? You and your colleagues have, over the past century, exposed the secrets of life on earth, sir. Thanks to you, we possess just about everything we require in order to modify every scrap of organic life, in whatever way we choose, as well as to create new life forms from scratch, where we find nature lacking in sufficient ingenuity."

"An interesting digression. But I'm still waiting for your answer to my question."

"I'm not trying to wriggle off the hook, as it were. Here's my short answer. We at the Sujana Foundation are of the opinion that the science of life is far too dangerous an implement to put into the hands of people who believe in the existence of demons, devils, and other insubstantial entities. The very same people who long fervently for the second coming of the Messiah or the reappearance of the Twelfth Imam, followed in quick succession by the End Times and the Last Hour, when an orgy of destruction signals that the game is up for sinful humankind."

"My goodness, Gaia. That strikes me as being an elementary flaw in logic. Our science isn't what's dangerous—it's those people who present the danger. Let me ..."

Gaia's hearty laugh interrupted him. "Where have I heard that before? Ah yes, I recall now that profound expression, 'guns don't kill people, people kill people.' The very model of a proposition that manages to be both trivially true and truly trivial simultaneously."

"Please enlighten me. Exactly in what way is it trivial?"

"Let me answer you with a parable. My sisters and I became so fond of this literary form during a curious period in our lives that I won't bother to explain. I call this 'the parable of the patriotic chemist.' The storyline itself is derived from the biography of a remarkable man named Fritz Haber, a German scientist of astonishing brilliance who lived from 1866 to 1934. His contributions to fundamental science were matched by an endless series of amazing technical innovations, and almost single-handedly at the beginning of the twentieth century, he forged the close and enduring connection between university science and the chemical industry.

"His greatest, earth-shaking discovery, for which he was awarded a Nobel Prize in 1918, was the synthesis of ammonia from air. Of course, you know that ammonia is the key ingredient in chemical fertilizers, an innovation that led to an explosion in human numbers by virtue of its impact on agricultural productivity. About half the world's population is alive today only because of it. Ammonia is also a vital component of high-explosive munitions, but that's another story, although it does suggest a fearful symmetry between the generation of more bodies on the one hand, and creation of more efficient means of blowing them into smithereens on the other."

At that precise moment Athena walked in. "I heard that. You're telling the story of Haber's life, if I'm not mistaken."

"Greetings, sister, your hearing is faultless. Let me introduce you to Professor al-Dini, one of our recent recruits." They nodded to each other. "I believe you've already made the acquaintance of the other character here. Sit down, let me finish my conversation. Now, where was I?"

"You were speaking about Haber's synthesis of ammonia and what innovations succeeded that triumph," al-Dini answered. "But I confess I do not see the point. If I'm not mistaken, a parable is a story to make an educational point."

"Bear with me a little longer and you will have it. German professors of the last century were civil servants, and none could have been a more dutiful servant of the state than Dr. Haber. During the First World War he was part of a team that developed chlorine gas as a chemical weapon. But this strange man was not content with confining his genius within the walls of a laboratory. He personally supervised the first release of the gas in warfare—against Allied troops stationed in their miserable trenches on the Western Front at Ypres. His grateful Kaiser decorated him for meritorious service and promoted him to the military rank of captain, even though he was too old to be a soldier.

"His wife, however, had a slightly different reaction. Clara Immerwahr, whose last name means 'always true,' had been the first woman to earn a Ph.D. in chemistry at the University of Breslau, the city—later Wrocław, after it became Polish territory—where she and Haber were born. Alas, the social conventions of the day dictated that she must give up her scientific career when she married. She discovered her husband's work on chlorine gas accidentally, whereupon she began to engage in bitter arguments with him, trying to dissuade him from weaponizing the material, to no avail. The couple gave a dinner party after his return from the Ypres front, to celebrate his promotion to captain. They had another furious row, she accusing him of perverting the mission of science. Later that evening she retrieved his service revolver from its holster and shot herself through the heart. When she died the next day, Haber asked others to oversee the funeral arrangements so that he could depart immediately for the Eastern Front, where he repeated the experiment against Russian soldiers."

"I trust you will soon draw some appropriate lessons for us from this sad tale, Gaia."

She hesitated and looked at him a bit crossly, then continued. "Actually, it gets worse. After the war, he commercialized an important class of insecticides and disinfectants made from hydrocyanic acid. The patent was held by I. G. Farben, the huge German chemical industry conglomerate. The product was known as Zyklon B."

I piped up, helpfully. "Wasn't that the gas used to kill people in the Nazi concentration camps?"

"Hush, Marco," Gaia said with a smile, "you're getting ahead of me. The unraveling of this triumphal career began in the simple fact that Haber was born a Jew, although he had converted to Christianity early in life. When Hitler came to power in Germany, he ordered the dismissal of all Jews from the civil service, and in Germany this included university professors. Jewish-born *conversos* such as Haber received no exemption, despite his having been such a faithful servant of Prussian militarism. His great good friend Einstein had seen the handwriting on the wall and was already out of the country. When in April 1933 the Prussian Academy of Sciences was ordered by the Nazi authorities to condemn Einstein publicly on account of his criticisms of the German government, Haber dutifully supported the motion. Less than a year later he was dead of heart failure.

"Now, Dr. al-Dini, I want you to help me draw the moral from this story. I will pose a hypothetical question for you. Assume if you will that the esteemed Herr Professor Doctor Haber had lived on in his native land into the early 1940s. Try to picture him as his stands together with a group of his relatives in the Nazi concentration camp at Dachau. Imagine him in your mind as he awaits the inception of the announced delousing routine in the gas chambers, and as the light of awareness switches on in his mind: The officials will, of course, be using Zyklon B, his wildly successful innovation. Here's my question. Do you see, in your mind's eye, a smile of satisfaction passing across his face as the cyanide gas begins to pour into the chamber? Now, as it hits his nostrils, would a feeling of deep inner joy suffuse his soul at this additional proof of how well he was helping his country again, one last time? Would his chest expand with pride at the realization that his gift had become the preferred instrument for the mass murder of his people by the state he had served so loyally?"

He glanced nervously about the room. I'm only guessing, but I'm pretty sure he had the distinct impression that he was being set up. So he played for time. "Forgive me, Gaia, but I don't understand your questions. The tribulations this man suffered are bad enough, without having to imagine additional blows raining down on him. It seems to me he deserves pity, or perhaps empathy, but not mockery."

"A fair comment, perhaps, but I swear I don't intend to deride him or mock the suffering he experienced at the end of his life. His is an extreme case, and I cite it here because such cases pose issues more clearly for the rest of us. I asked you to imagine what emotions might have coursed through his heart had he ended up in the same place where some of his relatives actually were murdered for a specific reason—because he himself had once stood on the other side of the apparatus, so to speak. At the onset of the Great War he had badgered a reluctant German military high command to use chlorine as a weapon. Having finally succeeded, he personally helped to release the deadly gas at Ypres in 1915, which killed or horribly maimed thousands of soldiers, and then he did it again on the Russian front. Those poor souls never knew what was happening to them, of course.

"But not so in Haber's case. In my imaginary scenario he would have known exactly what was going to happen to him and his companions, crammed in cheek-by-jowl on the floor of the extermination chamber. So I ask you again, put yourself in his place: What do you suppose would be his last thoughts about his science in the moments before the cyanide gas hit his nostrils?"

He was silent for some moments, and it was obvious he was peeved, but al-Dini's voice was even when he finally spoke. "I do see your purpose in this exercise, since I am aware that the objective of a parable is to express some general truth in the form of particular circumstances, as in the story of the loaves and fishes in the Christian Gospels. Yet you must pardon me for my obtuseness in this case. Try as I might, I cannot see the larger meaning you wish to convey in this tale, since the events you refer to in its closing scene never actually happened to this great and unfortunate man."

"It is my duty to apologize, Professor; I swear I'm not playing games with you in order to have some fun at your expense. Let me state the point plainly, then. Isn't it obvious that, for all Haber's genius, he was wretchedly naïve—quite literally, murderously naïve, I believe I can say without the slightest exaggeration—about the relation between his science and his society? Isn't it crystal-clear that the thought never once crossed his mind that he might bear some personal responsibility for the atrocious deeds done with his inventions? And, in the case of the chlorine gas he weaponized, done as a result of his persistent and heartfelt entreaties?"

Here Athena laughed softly. "You were right to apologize to our guest, Gaia. After all, you and I have a long history in this matter, years of discussions that culminated in one of the great battles between our father and Hera." She turned to al-Dini. "I heard much about you before your arrival among us, and it's good to meet you at last. This discussion is not really about Fritz Haber. It's about the relation between scientists and the products of their work in the context of the society that promotes and pays for their activity. To a man or woman, with the rarest of exceptions, scientists are wont to draw a firm line between the new knowledge generated in their labs and what is done with it in the wider world."

"Not so firm a line when it comes to patents and royalties, sister," Gaia remarked.

"True enough, but what I said is still a general truth, because the great majority among the millions of dutiful scientists never patented anything. Moreover, although after about 1850 most were salaried, they have shared a pure ethos of discovery, a belief that their new knowledge is both a good-in-itself—an act of enlightenment for the human species—and, in most cases at least, a benefit to their fellow humans, a promised betterment of the conditions of life."

"That is what I, too, believe, to the bottom of my heart," al-Dini interjected. "In my place of origin, the state claimed the right to patent and extract royalties. What did this matter to me? I never once considered challenging that right, or leaving my native country for greener pastures, as the saying goes. I felt I was born to be a scientist, and privileged to be able to join a community of researchers stretching centuries backward in time and forward indefinitely into the future. I was deeply honored to win the respect and recognition of my peers elsewhere in the world. What drove me to seek out my laboratory, day and night, was the challenge of confronting and solving a few of nature's clever little mysteries."

"Well said," Athena replied. "You sound exactly like my late and very distinguished father!" at which remark she and Gaia broke out laughing together. "And I'm glad you mentioned your membership in the transgenerational community of scientists, for this collectivity, and not individuals such as yourself, is my referent. For them, the distinction between their responsibility for their own discoveries, achieved over time by the contributions of countless hands and minds, by contrast with the applications that may flow from them, is their principal article of faith. Gaia chose the extreme case, in the person of Haber, to suggest to you that we, at least, are not believers in this regard."

Determined to keep up his side of the conversation, al-Dini said, "May I interject? I must concede his culpability in the case of chlorine gas, based on the story you have told me, the truth of which I have no reason to doubt. But how could Haber have anticipated the rise of Nazism, the program of wholesale extermination of ethnic groups, a program implemented with deliberate sadism and savagery? How could he have suspected that his insecticide, with its promise of great social benefit, would one day be used by a German state—a state he had quite obviously worshipped—against his own kin, and millions of others besides?"

"Gaia," Athena said, smiling, "I forbid you to speak in parables from now on." Then she turned back to their guest. "Of course, he couldn't have been expected to foresee such specific outcomes, Professor. But I do think that it should have been obvious to him, as he stared into the trenches along the enemy front, those filthy pits of hellish misery, and as he prepared to unleash his own satanic innovation, that his science had now revealed itself to be a ghastly, double-edged sword. The Great War was the turning point in modern history, in many ways, including the role that the modern sciences have been assigned to play in the drama. Haber was right there, right in the middle of it, the truth was staring him in the face, and he was utterly oblivious to what was happening."

"In what sense was the Great War a turning point?" al-Dina asked.

"In precisely this sense: The modern natural sciences were born of a noble impulse, the vision of an international community of like-minded searchers into nature's truths, the vision of a truth-telling form of knowledge free of religious fanaticism, ethnic divides, hierarchical social divisions, and the petty rivalries of nations. This remarkable community had grown into its self-appointed role over the course of the preceding three centuries, giving rise to something entirely new in the course of human history to date. And then that noble impulse was tortured and broken on the wheel in the nationalistic madness that engulfed the Western peoples in 1914."

I decided to intervene. "What exactly had happened up to that point, if I may inquire?"

"Seventeenth-century scientists began the practice of regular correspondence with their peers abroad. By the close of the nineteenth century, the national academies in various nations, which included most of the eminent scientists of the day, were regularly appointing foreign fellows to their ranks, solidifying the transnational research collaborations that had been developing. Then in 1914, in a matter of mere months, all the goodwill vanished. Some of the most distinguished German scientists and intellectuals had put their names to an 'Appeal to the Cultured World,' a document full of blatantly racist ideology; among the ninety-three signatories were Max Planck, Fritz Haber, Ernest Haeckel, and Wilhelm Roentgen. The signatories threw their support unreservedly behind the German government and its war effort, denying—without benefit of investigation—the reports of atrocities committed by the German army against citizens of Belgium, a nation that had proclaimed its neutrality. Scientists in other belligerent countries reacted with equal antipathy, and various academies in the warring nations began expelling their foreign fellows. The dream of a science beyond politics went up in the smoke of artillery barrages."

"If my memory serves me well, most of these relationships were restored quickly after the Great War," al-Dini stated.

"Indeed, many of them were, only to be shattered again, more brutally this time, little more than a decade later with the coming of fascism," Gaia responded. "If you're willing to continue these little chats, Professor, as I hope, I'd like to review that period with you as well, a period in which the new physics had become the groundbreaking science. But I want to go back to 1914 for a moment, back to the turning point to which my sister Athena referred a moment ago. An opportunity was lost there, one that would never be presented again. There was an alternative choice for these world-class scientists, both Germans and others, that they spurned, almost without a moment's thought: A chance to choose universalism over nationalism. A chance to rely on the solidarity of scientists and the values of scientific inquiry, against the madness into which European civilization was fast descending, first in 1914–1918, then in its authentic sequel barely two decades later. They appear to have dismissed the very possibility of making such a choice."

"All except one—one alone amongst the top rank of their kind," Athena interjected. "Albert Einstein."

"Yes. Isn't it a truly astonishing event? Einstein and a friend of his drafted a counter-manifesto, 'Appeal to the

Europeans,' seeking to remind their fellow scientists and intellectuals that their primary loyalty was owed to European civilization as a whole. It fell on deaf ears; not one other prominent scientist offered to sign it. The dream was over, although Einstein himself never gave up: He signed the 'Russell–Einstein Manifesto,' a protest against 'weapons of mass destruction,' two weeks before his death in 1955. So did a handful of other scientists. But the moment had passed."

This sisterly colloquy was interrupted by al-Dini. "What exactly was the opportunity they let slip? Gaia, I believe that you phrased it as the failure in 1914 to choose universalism over nationalism. That is a true enough account, on the historical record. But let us suppose the opposite was the case. What real difference could have been made? Do you honestly think that a group of leading scientists, however eminent, could have altered the course of the Great War? Or what followed in its train, the further descent into barbarism beginning two decades later?"

Athena responded, "Almost certainly not. My unhappiness lies in the simple fact that they didn't even try."

"Try to do what?"

Here she paused for what seemed like an eternity. Finally she answered, "To sever their science from its service to industry and the state."

Before he could say anything, Gaia interjected. "Dr. al-Dini, I could sympathize if you thought we've been a bit unfair to you. Athena's remark reflects the stage in our thinking at which we arrived after much reflection and debate among ourselves. So please indulge me for a little while longer. Let me lead you back in time for a closer look at the longer interaction between modern science and its social context. Will you take a few more minutes to accompany me?"

"I can hardly decline such a request from you, since I am conscious of the many kindnesses you have showered on me and my family since we arrived here. So please, carry on."

"Thank you. Let's go back to where it all started, in the late eighteenth century, to the period known as the French Enlightenment. This is when Antoine-Laurent de Lavoisier lived, the man known as the father of modern chemistry, when the engine of modern science began to be constructed in earnest. But I want to focus on another remarkable individual from that period, a man with a delightful name: Marie Jean Antoine Nicolas Caritat, Marquis de Condorcet. By a strange coincidence, Condorcet has exactly the same dates of birth and death as Lavoisier: 1743–1794. A brilliant mathematician, he was elected a fellow of the Academy of Sciences at the age of twenty-six. After the overthrow of the monarchy, he was an elected deputy in the Assembly, where he was a passionate champion of the rights of women and blacks; but like Lavoisier, he was condemned unjustly and murdered during the Reign of Terror.

"Condorcet's legacy is summed up in his book, *Sketch for a Historical Picture of the Progress of the Human Mind,* which was first published posthumously in 1795. This text is the clearest statement of the idea that the new scientific methods are not only important for the truer understanding of nature. Rather, their highest importance lies in the fact that they can and should also be diffused throughout society, by means of universal education, and that social policy and social institutions will be rendered more humane and just as a result. I recently re- read his book and made some notes that might be helpful here. If you like, I'll give you a copy of them."

NOTES ON CONDORCET'S PROGRESS OF THE HUMAN MIND (1795)

GAIA SUJANA

The process called "enlightenment" is founded on a way of thinking that instructs us "to admit only proven truths, to separate these truths from whatever as yet remained doubtful and uncertain, and to ignore whatever is and always will be impossible to know." The gradual extension of this method into the realm of "moral science," politics, and economics has enabled thinkers "to make almost as sure progress in these sciences as they had in the natural sciences." With these words Condorcet envisioned a future in which "the dissemination of enlightenment" would "one day include in its scope the whole of the human race." He continues:

> This metaphysical method became virtually a universal instrument. Men learnt to use it in order to perfect the methods of the physical sciences,... and it was extended to the examination of facts and to the rules of taste. Thus it was applied to all the various undertakings of the human understanding.... It is this new step in philosophy that has for ever imposed a barrier between mankind and the errors of its infancy, a barrier that should save it from relapsing into its former errors under the influence of new prejudices.

Condorcet has an interesting reason for suggesting that advances in the natural sciences are the original foundation for a broader social enlightenment. He remarks that "all errors in politics and morals are based on philosophical errors and these in turn are connected with scientific errors." What he is saying is that there is a connection between our conceptions of natural processes, on the one hand, and our understanding of society and individual behavior, on the other; I find this to be insightful. Once the "progress of the physical sciences" is launched, he claims, this "inexorable progress cannot be contemplated by men of enlightenment without their wishing to make the other sciences follow the same path. It offers them at every step a model to emulate."

This theme is nicely summed up in the following sentence: "Just as the mathematical and physical sciences tend to improve the arts that we use to satisfy our simplest needs, is it not also part of the necessary order of nature that the moral and political sciences should exercise a similar influence upon the motives that direct our feelings and our actions?"

If there is one core idea in Condorcet's conception, it is surely this: The "progress of the sciences" that defines the Enlightenment project is a double-sided phenomenon. It encompasses both the physical and the moral sciences or, using my terminology, the combination of inventive and transformative science, or technology and ethos. It is a process with a built-in mechanism ensuring its indefinite continuation: "The progress of the sciences ensures the progress of the art of education which in turn advances that of the sciences."

The inner unity between these two dimensions is something that Condorcet seems to have taken for granted. He saw the two sides as arising in quick succession over the course of the seventeenth century and flourishing together throughout the eighteenth. In short, a more sophisticated chemistry and physics on the one hand, and enlightened social behavior on the other, were two sides of the same coin. That this is an inner unity, and not just a coincidence, is shown by Condorcet's emphasis on the great advances made possible by the invention of the calculus: It is not only a methodological pillar of the new natural sciences, but also of such innovations in social welfare as insurance and pension programs, which require the use of probabilistic analysis in order to function well.

In short, Condorcet thought that the new science would utterly transform human societies in two complementary ways—in living standards and social behavior. He believed that its twin aspects—the physical sciences and the moral sciences—would continue to advance together in lockstep. The idea of the unity of the two is symbolized, for me, in the remarkable coincidence between the life spans of Lavoisier and Condorcet. But it failed to take root. Instead, what happened was, as the epoch of modernity unfolded in the period after 1800, the two sides did not in fact continue to support each other, but instead split apart.

This resulted in the *hyper-development* of one side (the physical sciences) and the *under-development* of the other (the moral sciences): Beginning with nineteenth-century industrialism, radical changes in the technologies of industrial production simply swamped the far weaker trends in social transformation.

What happened on the European continent during the first half of the twentieth century provides the grim proof of the failure of Condorcet's vision. The two world wars acted as a kind of one–two punch against the illusions summed up in the notion of the "progress of civilization." In the ideology of nineteenth-century imperialism, the mission of civilizing the savage peoples of Africa and Asia had been assigned—by themselves, of course—to the European nations. The First World War shattered this first illusion definitively, for how else could one describe what happened on the battlefields of Belgium and France in the years 1914–1918 except by the terms barbarism

> and savagery? And it was the introduction of gas warfare in 1915 that provided the first intimation that there was worse to come.
>
> By 1945, it had become abundantly clear that what had happened in Europe over the preceding thirty years was no "reversion" to a state of "primitive" murderousness. No, this was a march forward to a new state of affairs, where the most advanced products of human reason would be called to the service of the most depraved and basest impulses.

She looked up from her papers. "Now, I hope, the import of Athena's remark will be a bit clearer. It is not in the progress of science itself where the trouble lies. Rather, we see the ever-closer melding of science with industry and the state, which begins toward the end of the nineteenth century, as the source of our concerns. To be sure, science itself is propelled forward at an accelerating pace by this conjunction. *But social behaviors are not.* The Enlightenment idea of balanced progress in these two domains is left in the dust: Science became universal, but enlightenment never did. In a nutshell, we want to extract science from this fateful nexus until society has a chance to catch up."

He sat quietly for a moment before speaking. "What can I say? You've offered me much rich food for thought. I need some time in which to digest it."

"Then let's stop there and continue the conversation at another time, shall we? Right now, I have some other urgent business to attend to. Dr. al-Dini, I would very much like you to resume this discussion with me, if you're willing to do so; and I hope that Hera can join us at that time. She and I are in fact the founders of Solomon's House, and jointly we are responsible for its management, security, and financing. You are a very valuable new recruit for us. I want you to be comfortable with what we are doing as well as the reasons for the decisions we have taken regarding the dissemination of scientific knowledge. Will you do this for us?"

He said simply, "Of course." He rose, shook hands all around, and quickly exited.

"Athena, this is your first trip to Las Vegas in ages," Gaia remarked. "You don't get out very much these days! Let's walk around a bit and I'll show you what we've been doing."

"True enough, sister; both Hera and I have been focused on the 'summer-camp' part of our enterprise. Our youngsters are about to turn eighteen and they're a rambunctious lot."

As we stepped outside after visiting a few of the high-security lab facilities, the hot desert air that greeted us was eerily quiet. Sometimes it's hard to imagine that tourists once thronged the shops in the now-vacant casino malls and aimlessly filled the sidewalks amidst the twenty-four-hour flow of traffic along the Strip. Now, in the evenings, after sundown, I'm pretty sure that the coyotes busily afoot inside the city limits outnumber the few human residents who occasionally emerge briefly from their underground hideouts. Already, few traces of the original streetscape, molded from the hijacked spirits of departed cultures, remain visible. A hundred years from now only the urine-soaked detritus lovingly preserved in the pack-rats' middens will retell their tale. Future primate archeologists, having stumbled upon them a thousand years hence, then pondering the fragments of ancient Egyptian artifacts therein, will go mad trying to sort out the historical chronology.

"Did you take the train from Yucca?" Gaia asked.

"Of course; it's so fast."

I recalled the construction project I had supervised a few years ago, when we built the 100-mile secure rail corridor, flanked by pipelines for water, electricity, and fiber-optic cables, following Highway 95 from Beatty to Las Vegas and including a spur line to Nellis Air Force Base. Although the facilities for Solomon's House are located along the downtown corridor where the casinos and hotels once flourished, the façades of the grand hotels themselves are no longer visible, having been draped from top to bottom in huge sheets of tough, flexible black plastic that shield the buildings' exteriors and also serve as massive solar collection panels.

The residents live in the bottom three floors of the hotels, the upper stories of which are vacant. All of the labs and manufacturing facilities are underground, in the former shopping areas. Heat exchangers capture the cold desert night air for air conditioning and the daytime heat for warmth. Other facilities—classrooms, recreation areas and gyms, restaurants and commissaries—are in the adjacent low-rise buildings near the hotels. A security fence surrounds the downtown core. The former suburbs are deserted and are beginning their long, slow return to the earth.

My administrative burdens were lightened considerably once the Foundation's board of directors had approved Narciso's appointment as my second-in-command. Since among my duties was serving as chief of security for all our facilities in the town of Beatty, at the Yucca Mountain site, in the ruined cities of Las Vegas and Bakersfield, and at our Jalama Beach retreat on the Central California coast, we were soon referred to by the inhabitants thereof as "the sheriff" and "the deputy."

Shortly thereafter, the two of us and members of our security detail were standing at the rail junction in Bakersfield, California, while a large shipment of medical supplies was being unloaded into our trucks. The US Air Force maintained the regional rail system in order to provision its major installations in the southwest—linking Nellis Air Force Base in Las Vegas on the eastern end with Edwards AFB, located at the southern tip of the Sierra Nevada mountains, and continuing westward through Bakersfield to its terminus at Vandenberg AFB on the Pacific coast. Since we were partnered with them through our contract for the Yucca Mountain facility, and also had operations at both ends of the line, including our summer retreat and primate sanctuary at Jalama Beach, just south of Vandenberg, we often attached a group of our own rail cars to one of the military supply trains.

"When we arrived during the night," Ciso remarked, "I

noticed that only the train junction and that big complex in the distance had any lights burning."

"That's our Bakersfield Hospice in the distance, which used to be the state university campus here. As you'll see, it's huge, some two hundred buildings spread over a thousand acres, all surrounded by a security fence. We have one of the compact, helium-cooled nuclear reactors to generate electricity. There's no electrical grid left anywhere in the southwest, so if you want to run an operation you have to generate your own."

"So what's a hospice and why do you run one?"

"We call it that, rather than a hospital, because a lot of what we do is just the relief of suffering for both people and animals. It's the only public facility of its kind in the entire region. We provide a range of basic drugs and medical and dental services, along with painkillers, contraceptives, vitamins—you name it—all free of charge in the outpatient clinics. We also admit patients to carry out some simple surgical procedures, but nothing complicated; in many cases we simply try to keep people free of pain as they're dying. We operate it under contract with the US military administration in the southwest region and bill the costs to them. A big part of our contractual responsibility is to analyze blood samples to keep track of infectious diseases caused by viruses and bacteria, which are always mutating, and especially to quickly identify anything that looks like an engineered pathogen that was designed as a bioweapon. We provide vaccinations because we can administer them ourselves, but we don't hand out antibiotics, except to those who've been admitted to the facility."

"Why not?"

"Because people rarely follow the antibiotic regimen faithfully and so all you are doing is helping to breed a new generation of superbugs."

"Did you say it was the only facility of its type anywhere around here?"

"Yes. You have to remember that the folks who are still hanging on here represent a fraction of the total number as it was fifty years ago. There's very little water left and whatever's still available is untreated and carries various diseases."

"It's the same in Mexico," Ciso added, "as I saw when I made my way through it last year. At times it was touch and go for me. I had to follow the animals to the few remote springs scattered throughout the desert."

A few hours later, our convoy proceeded under guard through the favelas that ringed our hospice on all sides. The smells emanating from our biodiesel-fueled engines prompted some of the residents to gaze hopefully in our direction. "We run mobile soup kitchens for the residents whenever we have surplus food at the hospice. They must have mistaken us for the dinner wagons."

"Where does the food come from?"

"There are something like two dozen Hutterite colonies operating along the Central California coastal region within the extended security perimeter north and south of Vandenberg Air Force Base, including the colonies in the Jalama Valley, which is part of our territory. In return for the right to farm the very fertile land in that region, the colonies produce a huge quantity of extra food crops that is then transported here. When we've finished delivering these supplies to the hospice, we'll hop back on the train, which will be continuing on to Vandenberg. You'll be amazed at what you'll see in the Jalama Valley."

And he was. Our own tribe had just returned from there to Yucca Mountain by train after spending the summer months away from the fierce desert heat. We stood on the observatory platforms overlooking the western end of Jalama Valley, our backs to the nearby Pacific

Ocean, and with our high-powered binoculars, took in the activities of the menagerie below.

Our primate sanctuary, segregated into separate enclosures for gorillas, chimpanzees, orangutans, and bonobos, now numbered thousands of inhabitants and was still growing. For nutritional purposes, we had introduced plants that mimicked their native African diets. Along the security fence we allowed places of ingress for small local mammals to provide hunting opportunities and dietary protein for the chimps and bonobos. Apart from being fitted with miniature tracking devices and having their blood occasionally sampled by our keepers to watch for new viral outbreaks, they were left entirely alone. Observational research was carried out from the boundaries of the various territories using platforms, although a network of video and audio recording devices was also concealed in the trees inside.

After rechecking the security installations at the Jalama complex and holding one or two meetings with our counterparts at the nearby air force base, we hopped the next train back to Bakersfield. There we set out with our security detail in a smaller supply convoy, heading northeast and steadily gaining elevation, on the single road that winds for thirty miles along the lower Kern River Valley leading to Lake Isabella. This peaceful and secluded spot is where we built our Palliative Care Facility, which serves those of our own tribe and the staff complement from Yucca and Beatty who require long-term recuperative treatment or end-of-life care. Having completed our mission, we transited the Walker Pass and dropped down into the desert, heading through Death Valley back to Beatty.

7

The call on the satellite phone came at midnight, just after Narciso, Magnus and I had turned in after a long day of hiking. Hera's voice was calm and clear. "Marco, listen carefully, please. Two of our boys have kidnapped Lucetta." It was the first time in months that we had spoken with each other.

"They've taken a camouflaged truck and disabled the GPS unit, so we can't track them by satellite. Almost certainly they'll be heading for Las Vegas, where there are plenty of buildings for them to hide out in for a while. But we've blockaded all the paved highways, so they'll have to go by the back roads. Is there any way you can intercept them?"

"We're in Ash Meadows. This is the only alternative route to Vegas by road," I replied.

The fugitives would have reckoned that the only direct route—along Highway 95—would be controlled by checkpoints that would intercept them. I know the area map by heart: They would have to head for Las Vegas through Ash Meadows because there are no other convenient backcountry routes from the Beatty area to the old casino city. The route they would have to follow first heads south on Highway 373 and then cuts directly through the Ash Meadows wildlife refuge. Once you leave the highway, the secondary road—really a dirt track—comes to a "T" junction, where they would turn right and head toward Bell

Vista Road, which would take them through the now-deserted little town of Pahrump and then to Highway 160, leading into Las Vegas from the west.

"There's only one road through Ash Meadows, Hera. They'll be passing within a mile of our campsite at Point of Rocks Spring. When did they leave Yucca?"

"They crossed through the security gates five minutes ago, which, of course, set off our alarm systems."

"They haven't driven these roads before, and there are some tricky turns to navigate. I'm guessing we'll have close to an hour to set up our ambush. We've got to get moving. You can reach us on the sat phone. Bye!"

"Wait! Marco, they're armed with assault rifles from the armory. Stop them. Do whatever you have to do. You're authorized to use lethal force. I want Lucetta back in one piece. The other two have forfeited their right to the same consideration."

The gear we needed was already packed inside our own vehicle, so the three of us leapt in and roared away from the campsite under a brilliant moonlit sky. The unmarked track leading from the "T" junction to Bell Vista Road, affording easy access to a number of the best springs in the area, runs straight for some miles, which allows for a fair amount of speed. We needed to slow them down. First, we moved the sign for the turn to Pahrump, which usually sits at the intersection with Bell Vista Road, back up to a "Y" junction closer to our camp. Our plan, made on the fly as we raced down the road, was to divert them onto the smaller and bumpier road that goes past Jackrabbit Spring—in the vicinity of which there is some nice cover in low hills overlooking the road, on both sides, at a distance of about four hundred feet.

At the "Y" junction we set up another road sign from the collection I always carry with me in the area: "Danger: Washout Ahead. Detour." Often in winter the desert's notorious flash floods slash deep trenches across the roadways. Since I'm always traveling through this area, I monitor the damage and call in repair crews: This is a useful alternative route in case of emergencies on the main highway. A few yards further on past the junction, at a slight dip in the road, we threw down an ugly-looking spike strip designed to deflate a vehicle's tires. If they ignored the sign and roared through, they wouldn't get far. They would have to abandon the vehicle and then we could hunt them at our leisure using our trusty tracker dog.

We arrived at Jackrabbit Spring, concealed our vehicle, and unloaded the cases containing two of our sniper rifles and their thermal-imaging night telescopes. "Well, I guess it's time to put all of our practice with these things to use, my friend. You're the best shot, so take the Walther. Look around and set up the ambush. Tell me where you want me to be positioned."

Narciso surveyed the ground around us with his night vision binoculars, spotting the low rock outcroppings a short distance away on either side of the road. "I want you on the right-hand side of the road. Take the binoculars; from that rock outcrop up there you can see the 'Y' junction. Keep in touch by radio and let me know whether or not they make the turn at the phony detour sign."

We were standing together at a point on the road crisscrossed by deep ruts. "They'll have to slow down to a crawl right here. Use the rock just behind you to calibrate the distance with your laser. You're going to shoot out the right front tire." He looked at me and grinned. "Can you hit a truck tire at that distance?"

Our laughter broke the tension. "Good. Remember, do breathing exercises to stay as calm as possible. You'll get off a better shot that way. Hit the vehicle just before it reaches this spot. I'm pretty sure it'll swerve to the right and hit the rock. They'll probably assume they've dropped into a hole."

"I'll put the silencer on the muzzle of my rifle," I said. "That way they're not going to be able to hear the first shot or see the muzzle flash, even if they have the windows down, which they probably won't because of the dust. Where are you going to be?"

"I'm going to be on the left side, over there. Presumably someone will get out to see what the damage is. We need a plan for what happens next."

"Hera told me they took assault rifles with them. If anyone who exits the vehicle on your side is carrying a weapon, aim for it and try to disable it. I'll do the same. There's no silencer for the Walther, but that's fine because at that point I want them to know what's happening."

"You do know that unless we're extremely lucky they're going to get hurt, even if we both manage to strike the weapons with our shots."

"That can't be helped. They have a hostage. We need to end it quickly." I dropped a small radio transmitter behind the rock. "We may be able to pick up their voices with this."

I slung my weapon over my shoulder and headed for the matching rock outcropping on my side to set it up on its firing stand. In response to my command Magnus got up and walked beside me. He was outfitted in the desert gear I had made for him, the leather vest and set of booties for his feet, to protect him from the abundant thorns and cactus spines that lay about everywhere.

Ten minutes later Magnus perked up his ears. I radioed Narciso. "Magnus has heard something, although I haven't. It may be the vehicle." I had my binoculars trained on the junction. Then: "They've arrived." A pause. "They've made the turn and are headed this way!" I looked at Magnus and repeated the "stay" command.

Narciso heard the excitement in my voice. "Do your breathing exercises, Marco."

A few moments later, the truck hove into view, moving cautiously along the deeply rutted roadbed. As they reached the target spot I squeezed the trigger; no sooner had the muffled crack of my rifle sounded than I saw the vehicle lurch violently to a halt. I sighted the front passenger door through my scope.

A moment later a figure hopped out on my side, holding a flashlight but no weapon. A second figure emerged from the driver's door; slamming the door shut, he moved quickly toward the front of the vehicle, whereupon I could see that he was hefting a rifle. The sounds of their voices carried across the still desert air. The one with the light had seen the bullet's entry hole in the tire. The driver wheeled toward the darkness around him, holding the gun to his shoulder and moving it in an arc across his body, firing blindly. At that instant Narciso's single shot rang out. The weapon shattered in the man's hands and he let out a terrible scream as he dropped to the ground.

The second figure emerged from behind the vehicle in full flight. I commanded: "Magnus, go!" The dog took off like a rocket, covering the distance between them in seconds. I saw the running figure glance quickly over his shoulder in the instant before Magnus flew through the air and struck him from behind with his front paws squarely in the middle of his back. He was propelled face-first to the ground as Magnus recovered from the impact and scrambled back, now barking and snarling to immobilize his prey, as the training routine had directed.

"Ciso, head for Magnus and pick up that one, then meet me at the vehicle," I yelled into the radio. "I don't think he has a weapon, but be careful."

I left my rifle and ran toward the screaming man. When I got there he was holding both hands to his bloodied face, yelling, "My eyes! I can't see!" I ignored him and yanked open the back door. Glancing inside, I saw Lucetta huddled on the seat, handcuffed to the side of the vehicle, shivering with fear and suppressed rage, unable to move.

As I slid next to her in the back seat and held her, she cried out, "Oh Marco, thank God it's you." I kissed her and asked, "Do you know where the key is?"

"In his pocket, I think."

"Will you be all right here for a moment? I'd better look at his wounds."

"Yes, of course, you must."

"By the way, are they who I think they are?"

"Rainer and Kenji."

I got back out with the emergency medical kit all of our vehicles carry. He was in a sitting position, quiet now except for low moans, and I knelt down beside him. "How bad is it?"

"My face. My eyes. I can't see anything." He pulled his hands away.

The damage was obvious. Narciso's bullet had shattered the mid-section of his rifle, sending razor-sharp shards of plastic and steel across his entire face. "There's not much bleeding, and not a lot I can do for you. I'll give you some shots and bandage your eyes. We'll call in a helicopter; it'll have a doctor on board."

At that instant Narciso and Magnus arrived, escorting their prisoner, who was groaning audibly—understandably so, since his face and hands had been liberally punctured by cactus spines. I turned back to the wounded man. "Where are the keys for the handcuffs?"

"In my left front pocket," he murmured.

"Ciso, retrieve the keys from his pocket and free Lucetta. Then use the cuffs on your prisoner. Get our vehicle and dump him inside." I knew that the other man wouldn't make the slightest move, since Magnus was guarding him, the dog's eyes locked onto his and every muscle tensed for trouble.

She came up behind me as I was tending to the wounds. She knelt and hugged me, kissing my neck, then took the sat phone from my belt. “I’ll call for a helicopter and medical team,” she said.

The Priesthood of Science

I had informed Gaia that I was eager to attend her next meeting with Abdullah al-Dini, at which they were scheduled to resume their earlier colloquy about science and society. I am so burdened with administrative matters these days that I rarely have a decent excuse to dine on intellectual fare; in addition, my main source of such nourishment formerly—Hera's private discourses—had dried up since she had secluded herself after the devastating experience of the kidnapping episode. Yesterday Gaia had called me with the news that the gathering would occur a day hence, and she requested that I come a bit early.

"I want you to quickly read through a memo I dashed off during the last week, Marco. It provides some background for the chat with al-Dini. You're aware that I'm fond of addressing complex issues by looking for a concrete setting in which to ground them—preferably, if possible, a biography of someone who actually played a role in events relevant to those issues. So I looked around and, lo and behold, I found a 'second Fritz,' another twentieth-century German scientist who shared that moniker with Haber. This one was a nuclear physicist named Friedrich Georg Houtermans who had the most extraordinary life and career. As the setting for our discussion, I'm going to give al-Dini only a brief synopsis of this biography because I think he'd be bored with having to listen to the whole story. But I thought you might be interested in the longer account."

I was indeed.

MEMO ON THE LIFE AND TIMES OF FRIEDRICH GEORG (FRITZ) HOUTERMANS
GAIA SUJANA

Houtermans was born in 1903 in Danzig, which at that time was still a part of Germany (later Gdańsk, Poland). The son of a well-to-do Dutch banker, he was raised in Vienna by his mother, Elsa Wanek, who was the first woman ever to earn a doctorate in chemistry at the University of Vienna. He became an exemplar of the carefree Viennese intellectual, a witty raconteur who preferred cafés as a place of work. At the age of eighteen, he went back to Germany, to Göttingen, to study physics under James Franck.

And what a brilliant scene from a now-vanished phase of European civilization that was! It was a time for the constellation of a new science—atomic physics—to glow brightly through the work of a phalanx of gifted and boldly adventurous spirits whose collective creativity would never again be equaled in that discipline. This was one of those great moments in culture, where all at once, seemingly serendipitously, geniuses cluster to create a radically new understanding of mind and nature. It happened in ancient Greece, with philosophy, geometry, medicine, and literature; in the Italian Renaissance, with architecture, painting, and sculpture, but also in science and technology; around the turn of the nineteenth century, in music, in Vienna, with Haydn, Mozart, Beethoven, and Schubert; and again in late-nineteenth-century painting, centered in France.

Another legendary physicist at the University of Göttingen, Max Born, was the second member of the double-star around whom this circle coalesced. In addition to Franck and Born, among those who either studied there during the 1920s, or spent a good deal of time there in those years on visits from other European universities, were Enrico Fermi, Wolfgang Pauli, Werner Heisenberg, Victor Weisskopf, Rudolph Peierls, Paul Dirac, Linus Pauling, Edward Condon, Samuel Goudsmit, Robert Oppenheimer, and the Russian George Gamow. Many of those listed here would later become Nobel laureates, although Pauling was the only person ever to win an unshared prize twice—for chemistry and for peace. Many of them were also more than just physicists.

Max Born himself was a deeply cultured man who set the tone for this group. During their lengthy evening soirées at the cafés, the animated conversations were as likely to be about Goethe's *Faust* as anything else. One story must suffice here. Arguably, the most dashing personality in the group belonged to Oppenheimer, who had an amazing facility for languages. One night, the Dutchmen Goudsmit and his collaborator, George Uhlenbeck, who later became known for discovering electron spin, were sitting in one of the cafés reading Dante's *Divine Comedy* in the original Italian—a language that Oppenheimer lacked. For the next month he was nowhere to be seen during the nightly revelry. When he returned to the festivities a month later, he was fluent in Italian.

With his raffish Viennese background, Houtermans fitted in well. During his stay in Göttingen he began the work that would make him famous—the discovery that the energy output of stars is derived from thermonuclear reactions. He completed his doctorate, married another physicist (Charlotte Riefenstahl), and moved to the Technical University in Berlin to work with Gustav Hertz, who had shared a Nobel Prize with James Franck. The year was 1932. With the Nazis' accession to power in 1933, he immediately fled to England, not because he was part-Jewish on his mother's side, but because he was a member of the Communist Party!

Two years later his political idealism prompted him to make a most extraordinary decision: He accepted a proposal to move, with his wife and two young children, to the nuclear research facility at Kharkov in the Ukraine. But by 1937 Stalin's horrendous purges were well underway, during which perhaps more than ten million souls were arrested and either murdered or sent to die in the labor camps of Siberia. Among the targeted groups were resident foreigners. Houtermans was arrested and imprisoned, although his wife and two small children were allowed to leave Russia, undertaking a dangerous odyssey that ended only when they later found refuge in the United States.

Long afterward, Houtermans co-authored with a Russian historian (both of whom used a pseudonym) what is surely the most extraordinary book ever written by a nuclear physicist—*Russian Purge and the Extraction of Confession*, published in German and in English translation (1951). There the two men describe in agonizing detail the process whereby the victims of the purge were required to invent purely fictitious "legends" detailing their own guilt (and naming others who were implicated in their acts) as the only condition under which the protracted interrogations, sometimes involving severe beatings, would cease. Thereafter, conviction and either immediate execution, or the slower form of death involving transport to the labor camps, would follow. Often the interrogators themselves, as well as members of the dreaded secret police—the NKVD—who had made the initial arrests, would show up among the newly detained prisoners. Of all the great writers in the early twentieth century, only Franz Kafka had anticipated the combination of pervasive bureaucratic anonymity, the wholly insubstantial basis of an individual's guilt, and the collective madness that bore fruit in Stalin's gulag.

Believe it or not, the story worsens! In August 1939, Germany and Russia signed the Hitler–Stalin Pact, one of the terms of which provided for the exchange of prisoners held by the secret police in both countries, whereupon Houtermans wound up back in Germany once again, now in the Gestapo's clutches. By great good fortune he was released a year later due to the intervention of another Nobel Prize-winning physicist, Max von Laue, who found him a job in an industrial scientific laboratory in Berlin, where he was set to work on the problem of nuclear fission. (Another very interesting man, the privately wealthy Manfred Baron von Ardenne, owned this lab; he held something like 600 patents in electronics, was among the inventors of television, and after the war, worked on the development of Russia's atom bomb.) Thus, Houtermans was drawn into the circle of those in wartime Germany, led by Werner Heisenberg and Carl von Weizsäcker, who were tasked with exploring the feasibility of an atomic bomb.

Houtermans was a man of great courage. After all the personal horrors he had already endured, he took another tremendous risk. Fritz Reiche (the third Fritz!) was a German-Jewish physicist who had received his doctorate under Max Planck, and who had collaborated with Haber on gas warfare during the Great War. Do you see how these German scientific biographies overlap?! Thereafter Reiche became a professor of theoretical physics at the University of Breslau, Haber's alma mater, until in 1934 he was dismissed from his university post, again like Haber, under the terms of the Nazi law forbidding Jews to hold government jobs or work in the professions.

Incredibly, it took Reiche until March 1941, just months before the Nazis forbade further Jewish emigration (in preparation for mass murder), to receive an exit visa for the United States. Days before his departure he was visited by Houtermans, whom he had known slightly, who gave him this message for his old colleagues in nuclear physics who had earlier escaped to the United States: "We are trying hard here, including Heisenberg, to hinder the idea of making a [atomic] bomb. Heisenberg will not be able to withstand much longer. Say to them they should accelerate if they have already begun."

Houtermans' message was conveyed by Reiche to Enrico Fermi and others in America. Eventually it reached the circle around Oppenheimer, who would soon become the scientific director of the atom bomb project. After the war ended, Houtermans made his way back to Göttingen, and in 1952 was appointed a professor at the University of Bern. His earlier marriage had crumbled after a brief reunion and he remarried. A lifelong smoker, he died of lung cancer in Switzerland in 1966 at the age of sixty-three.

Seated in Gaia's elegant top-floor office in downtown Las Vegas, furnished with mementos of our days in the Turks and Caicos Islands, and with a splendid view of the Spring Mountains to the west of the city, I had just finished reading her memo when Hera entered. The anxieties induced by her recent experiences told in her face, but she had a warm embrace for each of us.

"I confess to you both I had to force myself to make the trip today, but based on what you told me of the first meeting with Professor al-Dini, I did not want to miss this one. I'll just listen today, please. I don't have the energy or wits for sharp dialogue."

Gaia just nodded, and I could see in her eyes the concern for her sister's well being. Immediately thereafter, al-Dini walked in.

"Welcome, Abdullah," Gaia said as she rose to greet him and to make the introductions. "Thank you for agreeing to continue our conversation. You already know that I'm inclined to focus on the lives and careers of real scientists when posing questions about the place of science in modern society. I found another interesting biography. Let me summarize it for you."

Before she could begin, al-Dini interrupted. "I've been thinking about the case of Haber, Gaia, and also doing a bit of reading of my own on it. And I agree there is something unusual about him, especially his determination to become so personally involved in persuading the army and government of his day to develop the technology of gas warfare, and most especially about his insistence that he be personally involved in the first uses of this terrible weapon. But in these respects he is—among the leading scientists of his day—the exception, rather than the rule, no? And if you agree with this statement, as I assume you must, how is it possible to draw any general conclusions about the scientific enterprise itself from this anomalous case?"

"You're right. I don't disagree with your assessment. My point isn't really about Haber as an individual. Rather, for me he is a symbol. Focusing on his life is a way of catching a glimpse of an underlying reality, which I would summarize in more abstract language as follows. On a purely conceptual and, simultaneously, intensely personal level, Haber's science belonged to him and his peers as their own path-breaking intellectual achievements. But the world around them was changing, as indeed it had started to do in the second half of the nineteenth century. I chose Haber as my example precisely because he was a chemist, and it was his field that first became 'industrialized.'

Modern science was entering a new phase where a reciprocal set of relations appeared—discovery, invention, and technological application, followed by commercialization, begetting a demand for more of the same, and attracting enlarged resources and facilities to ensure the recruitment of more scientists to satisfy that demand."

"Yes, and why not?" he replied. "New products designed for the betterment of life began to pour from the factories, but their ultimate source was the scientific discoveries made by Haber and his contemporaries. The synergies between the pure scientist and the commercial enterprise had reached a takeoff point, beyond which each would start feeding the other, endlessly. Haber may have first realized the synthesis of ammonia using nitrogen from air, but a subsequent scaling-up to large quantities was required before commercial production of fertilizer became a reality."

"Indeed," she answered quickly, "yet the first large-scale application was not for making fertilizer to produce more food, but rather for turning out high-explosive munitions during the First World War."

"True enough, Gaia, but was that Haber's fault—or the fault of Carl Bosch, the man who had figured out how to scale-up the ammonia reaction? There is simply no relation between, on the one hand, the motivating impulse that lay behind this great discovery, and on the other, the bodies of the unfortunate soldiers, choked on poison gas, that littered the battlefields."

"Not directly, no, but in a way you've made my point for me, Abdullah. The crucial point is that Haber's science no longer belonged to him or his peers—not in the way it had belonged to their predecessors, I suggest, in the period up until about 1850. The change was quite subtle, actually, hardly noticeable to the practitioners, because what mattered most to them, namely, science's *method,* remained firmly within their control. By that time scientists no longer had to answer to anyone else, such as theologians, who were speaking from a standpoint outside the borders of science itself—with a few exceptions, notably in the case of evolutionary biology, where the challenge from creationists remained lively. What I'm trying to say is this: Before society became enamored of the products of scientific discovery, the practice of science belonged wholly to the international community of scientists. Gradually, piece by piece, they've surrendered it to their paymasters."

"To speak frankly, I simply don't know what you're talking about, Gaia. And now I can return the favor and claim that you have made *my* point for me. What has always mattered most to all scientists is their method and the fact that no one but they control it. That remains as true today as it did in 1850, to use the dateline you proposed. From then to the present day, we have surrendered nothing of what really counts to the practice of science to anyone external to our community."

Gaia paused for some moments, looking vexed. I remembered sitting through some very similar conversations between Hera and her father, years ago, when I was a tenant in her beautiful house and gardens, built in the Japanese style, in the Caribbean town of Providenciales. Finally, Gaia spoke again.

"You are right to object because I appear to have lost the thread of my own argument. To be sure, science's method is the constant. Of course, it has been extended and deepened, but its essentials remain the same—a productive nexus of theory, imagination, observation, calibration, measurement, inference, new instrumentation, prediction, controlled experimentation, collaboration, peer critique, replication (or the failure to replicate), over and over again. Above all, it is the gradual accumulation of results, first by asking familiar questions and then by opening up previously unknown fields of inquiry, often gaining new insights that appear to have no practical benefits. But then, around 1850, the situation changes dramatically. The coming of industrial technologies provides a new avenue of application for scientific discovery and a whole set of positive feedback loops kick in—new science feeds new technology, which feeds new industry, leading to demands for new discoveries."

"Fine," al-Dini interjected. "You've now returned to the point of takeoff we had reached a few minutes ago. Then you derived a conclusion I disagreed with, something about science losing control over its own destiny at that point. Do you wish to pursue a different line of argument instead?"

"No, not really, but I admit I have to make it more persuasive. It's the accumulation of results that makes a difference, I think. This is what changes the situation, even though, as you rightly say, the underlying scientific method continues to strengthen along its original lines. Now the dominant socio-economic echelons in society sit up and take notice, recognizing that they might have underestimated the importance of what those strange scientists, huddled over their lab benches and breathing noxious fumes, have been doing all these years. Naturally, the scientists are flattered by this increased attention, along with which comes much more money and prestige. The scene is set."

"For what? Why does the accumulation of results itself matter? How does it change the situation of science, except for the better?"

"By Haber's time science was in a position to begin handing over technologies of increasing power, for which society sets about finding uses. And here's the rub: Those societies themselves hadn't changed all that much in the period since, say, the second half of the eighteenth century, when Lavoisier had begun systematically laying the foundations of this new science. True, he also improved gunpowder a bit for his country, which ultimately rewarded him for his trouble by beheading him in 1794 during the Reign of Terror. Asked to spare his life on account of his great scientific reputation, the presiding judge replied: 'The Republic has no need of scientists.' In a perverse sense that was a true statement for its time.

"Now fast forward about a century and a half in time. In 1933, Max Planck visited Hitler on a desperate mission to preserve some tattered remains of the extraordinary scientific community he had spent his life building up in Berlin. Seeking to persuade Hitler not to dismiss Fritz Haber from the Kaiser Wilhelm Institute, and clearly grasping at straws, he settled on the only argument he believed might resonate with the mad dictator, namely, that there was an important difference between 'productive' and 'unproductive' Jews. Quickly ending the conversation, Hitler replied, 'A Jew is a Jew.' Hitler was enraged that Einstein had managed to submit his resignation to the Prussian Academy of Sciences before the government-ordered expulsion order had gone into effect. Planck, a great and dignified man whose life was marked by a series of family tragedies, slunk away as Hitler launched into one of his pathological rages."

"Gaia, if you will permit me to be frank, your wonderful historical references are always most enlightening, but they rarely seem to serve the requirements of a logical argument."

She laughed. "I will focus on the compliment rather than the complaint, without faulting you for either. In the one hundred and fifty years that separate Lavoisier's fate from Haber's, their science of chemistry had revolutionized the human appropriation of nature's innate powers. But the society they both served so faithfully had undergone no such transformation in the same period of time! *That's* my point! In earlier times, the decadent form of rule by landed nobility and repressive church eventually incited spasms of murderous hatred during the Terror. During Haber's lifetime the remnants of older ruling elites joined with capitalist industrialists to win popular support for unleashing mass slaughter on the European continent, not once but twice in the span of thirty years. What had changed? For one thing, the number of victims had swelled by orders of magnitude, thanks to the greatly improved efficiency of the killing machines wielded by the state. The construction of those modern machines of death, including the Nazi extermination camps, would not have been possible without the new science of chemistry."

"I'm sorry, Gaia, I cannot see how you could lay blame for those dreadful events upon those who sought to unravel nature's mysteries for the good of humanity. I have already conceded Haber's special culpability, but you have provided not one other instance of personal involvement by a leading scientist in operating the machinery of death."

She sighed audibly. "I realize that my statements are often highly charged with emotion; it's one of my faults. Honestly, I don't attach blame to the scientists themselves, except for Haber, who—had he lived a generation later—would have been tried as a war criminal. What I'm trying to establish is that the great scientists didn't appear to see anything at all problematic in the relation between, on the one hand, their accumulating discoveries of new knowledge, and the technologies based on it, through which energies and materials of great power were harnessed, and on the other, the deep, structural irrationalities within the social institutions that almost certainly would misuse them."

Now it was al-Dini's turn to pause while formulating his reply. "That is indeed a preferable way of phrasing the problem you are trying to diagnose. It focuses not on the scientists' search for knowledge and truth as such, but rather on the relation between what the scientist discovers and what society *might* do with those results someday. But I'm sorry to say, although your point is phrased more appropriately, I don't think it's any more persuasive. For example, how long into the future would any individual scientist have to postpone his research before being assured that no type of harm would ever come from his findings? What if he or she reached the point of death without ever having found such assurance?"

Gaia smiled and slapped her desk with her palm. "Very well put! I'm so delighted that you agreed to join me in these forays, Abdullah. My thoughts become clearer because of how precisely you articulate your own responses to them. But I do have an answer to your questions. No scientist should ever feel an obligation to delay his or her search for truth, not even for a single minute. On the other hand, every scientist must take some personal responsibility for the outcomes in the larger social world that may reasonably be forecast." She held up her hand. "Before you ask me to explain what I mean by that statement, I'm going to request your indulgence once more. Please listen to a short summary of another scientific biography. I believe it will assist me to build my case."

He nodded his assent, and so Gaia turned to me. "Marco, since you've just read the little paper I prepared, would you summarize the highlights for us?"

I was just about to comply when Athena walked in and exchanged greetings with everyone. "Sorry, I meant to arrive earlier but was detained. Carry on, please." I then recited my lines.

"Astonishing, simply astonishing," al-Dini remarked immediately after I'd finished. "I confess I was utterly unaware of this Houtermans, and I doubt than many others have heard of him, at least outside the community of nuclear physicists. And what an extraordinary line-up of geniuses there was around Born in that field during those decades! Thanks to them, a mere twenty-five years following the ground-breaking work by Einstein and Planck at the very beginning of the twentieth century, all of the main foundations of an entirely new branch of science had been laid."

"Again, well said, Abdullah. I will simply add this observation: During exactly that same period of time, and in exactly the same location on the globe, another set of foundations were being laid, upon which was played out the greatest tragedy ever to unfold in the history of European civilization."

"Surely that is the merest coincidence, Gaia."

"I know you would like to think so, and perhaps it is. But the story of atomic physics will support my basic point. As is remarked in my little tale about the life and career of Fritz Houtermans, many of these geniuses were deeply cultured men and women. Now, in your mind, jump to the decade of the 1930s and watch a large group of them fleeing for their very lives from the European continent where they originated. Why? For no other reason than the fact of the ethnicity and religion of the families into which they happened to be born! How can we come to terms with the fact of such utter madness, and the unspeakable crimes to which it gave rise, occurring in what had become an age of advanced scientific knowledge? The change between Haber's time and theirs had stripped the veil from the comfortable belief in progress and enlightenment. The difference between Haber and those who were on the run from the Nazis was, at least for many of them, that a sense of dread about the future was now inescapable. With one notable exception—Einstein, who had already arrived at a similar state of mind during the First World War."

"I'm pretty sure I know where you'll be going with this story, namely, to the discovery of the nuclear fission reaction and the construction of the atomic bomb. But you have not addressed my point: The fact that these two sets of events—the rise of nuclear physics and the rise of Nazism—occurred at the same moment in time is a mere coincidence."

"Hear me out. Let's first finish the story of the discovery of nuclear fission." She shuffled the papers on her desk, scanning quickly the notes she had prepared for this meeting. "Here it is. Lise Meitner was the second woman to be awarded a doctorate in physics at the University of Vienna. She moved to the University of Berlin for further study with Planck, and in 1926 was the first woman ever to attain the rank of full professor, in any field, at that university. There she began a long collaboration with the chemist Otto Hahn, but in 1938 she fled for her life, ending up in Sweden. Born a Jew, she had converted to Lutheranism as a child, which, of course, made no difference at all in the eyes of the Nazi exterminators.

"She was the first scientist to realize that the nucleus of an atom could be split, and it was she who coined the phrase nuclear fission—with a little help from her nephew, Otto Frisch. The experimental proof was carried out in Hahn's laboratory in Berlin, after her forced departure; Meitner and Hahn separately published the results at the beginning of 1939. The group of nuclear scientists who had fled Europe and found refuge in the United States recognized immediately, of course, the implications of this proof. By the way, Meitner was cheated out of the 1944 Nobel Prize awarded for this work, which went instead to Hahn alone—not the last time this type of insult was bestowed on a woman scientist, as you may know."

"Another wonderful story drawn from the history of science," al-Dini muttered, "without providing any additional insight into the topic of our conversation, unless I'm mistaken."

I had the advantage of having known my aunt Gaia a lot longer than al-Dini had, and so I was well aware of her love of scientific biographies. I knew that the conclusions she had drawn for herself, about the general relation between science and society, had been formed in large part by what she had learned about the particular fates of eminent scientists during the first half of the twentieth century. A long time ago she had arrived at the personal judgment that the connection between what was happening in European societies between the end of the First World War and the onset of the Second, and the astonishing achievements of nuclear physicists during that very same period, was meaningful. But she also just loved the biographical record for its own sake, especially the episodes where women first earned their spurs in what had been formerly an exclusively male club. So I wasn't at all surprised to detect either the barely concealed look of annoyance on her face or the clipped tone in her voice when she spoke.

"Since you evidently enjoy them so much, I'll offer you one more, which I promise will be the very last." Again she scanned the notes on her desk, then picked up a new sheet of paper. "Leó Szilárd was a Hungarian-Jewish physicist who received a doctorate in Berlin in 1923 for a thesis that was extravagantly praised by Einstein. He then went to work at the University of Berlin with Max von Laue—you remember him from my other story—until he, too, had to flee in 1933. Shortly after arriving in London, he conceived the idea of a nuclear reactor, well before nuclear fission had been demonstrated, and later, once he had made it to safety in the United States, he patented a design for it with Enrico Fermi.

"As soon as Szilárd heard about the Meitner–Hahn results, he and Fermi, who were together at Columbia

University in New York, experimented with uranium and proved that a chain reaction was possible. Once the experiment was concluded, he recorded the following comment in his diary: 'We turned the switch, saw the flashes, watched for ten minutes, then switched everything off and went home. That night I knew the world was headed for sorrow.' By the way, he's the one who drafted the famous 1939 letter that Einstein sent to President Roosevelt. In 1942 he and Fermi, now at the University of Chicago, constructed the first self-sustaining nuclear fission reaction."

She looked directly into al-Dini's eyes. "Now do you see where I'm going with this?"

He replied, calmly and without rancor, but with just the slightest hint of exasperation in his voice, "No, I'm afraid not."

"That's fine, I've just set the stage for you, and here's the result: Those scientists, all of whom were collectively responsible for this breathtaking new knowledge, had lost control of its further development. At the instant when Meitner and Hahn delivered their two articles to the scientific journals for publication, the die was cast. The reason lies in the social context, that is, the imminent war to the death between opposing nations. Right away, at least some of those scientists realized that there would be a race to build an atomic weapon, and they also knew that one of the competitors for the prize would be a regime whose very political foundations were composed of mass murder, ethnic hatred, and a lust for world domination.

"The experimental proof that nuclear fission was possible was carried out in Berlin. A fair number of senior physicists, notably Werner Heisenberg, of course, were still there. Moreover, Niels Bohr, Heisenberg's teacher, was still in Copenhagen, and he only managed to escape with his family to Sweden and the United States in 1943, literally days before the Gestapo planned to round up him and the rest of Denmark's Jews. It was Bohr who had first bombarded uranium atoms in the search for the fission reaction, and who had written to Meitner and Hahn about this in 1938, which was the direct inspiration for the successful experiment they designed shortly thereafter."

"And yet what the others feared didn't happen, Gaia. Germany never managed to build a bomb, probably because Heisenberg put obstacles in the way, as your hero, Houtermans, clearly stated in the message he asked Reiche to deliver to his colleagues in America. For whatever reason, that aspect of the war had a happy ending."

"There's always been a fair amount of debate about what happened to the atom bomb project in wartime Germany, and especially about Heisenberg's role in it, as far as I know," she replied. "Yet how can we focus on how things happened to turn out, as opposed to the other outcomes that had a good chance of coming to fruition? Just consider how very close the world came to utter catastrophe in those dark days, Abdullah! Just imagine that some of the saner but still loyal Nazis, such as Albert Speer, had managed to wangle permission to assign high priority to an atom bomb project in, say, early 1940. Because it was in February 1940 that Heisenberg had completed, for the Director of the Nuclear Physics Section of Germany's Army Ordnance, his report entitled, 'The possibility of technical acquisition of energy from uranium fission.'

"Suppose they said to Heisenberg and the others, would you like to work hard on making this bomb project succeed, or should we send you and your families to the concentration camps instead? Germany had access to uranium, heavy water, and all other necessary resources to carry out the project. What would the world have been like with the bomb in Hitler's possession? And in the hands of his ally, Imperial Japan, already the author of horrendous war crimes throughout all of Asia? If all that had come to pass, as it might very well have, would we be sitting here today talking about the future of science?"

"Impossible to say, but I'm sure we're all glad things didn't turn out that way."

She seemed to be no longer paying attention to al-Dini's curt responses. "Some of the others, who still lived in free countries, tried to influence events. Meitner refused an invitation to go the United States, saying she would not work on a bomb project. Oppenheimer persuaded Bohr to visit Roosevelt, to see if he could win the President over to the idea of sharing the results of the Manhattan Project with the Russians. Roosevelt suggested that Bohr try to get Churchill's assent, and he tried to do so, but failed.

"James Franck—having fled Göttingen in 1933 for the United States—chaired a 'Committee on Political and Social Problems' associated with the atomic bomb. In June 1945, two months before the attacks on Hiroshima and Nagasaki, he and his associates proposed that the United States detonate a bomb on a barren island with an audience consisting of representatives of every country in the United Nations, as a way of showing the world what everyone was now facing. That July, in fact, the day after the first successful test of the bomb, Szilárd circulated a petition addressed to President Truman, which he persuaded dozens of scientists at Chicago, Oak Ridge, and Los Alamos to sign, that forcefully raised the issue of the 'moral responsibilities' incumbent upon those who would use such a weapon. It's highly likely that government officials made sure that Truman never saw it. Szilárd, Franck, and their colleagues were unsuccessful in all these attempts to persuade the US government to try other tactics before deciding to drop the bomb on Japanese cities."

Half-raising himself from his chair, animated and re-energized now, al-Dini leapt back into the conversation. "You see? Clearly, many of those most directly involved were acutely aware of their responsibilities in this grave matter, and even if they failed, they did make a determined effort to act according to their principles on that basis. You cannot deny the import of the evidence you yourself have sought to marshal against me."

Gaia had somehow rediscovered her normal merry mood and smiled at him. "Please, Abdullah, don't put it that way. I am first of all grateful to you because you've shown that you have listened carefully to my words. We can now both agree, after my lengthy recital of evidence, once and for all, that I have not come here on this occasion to condemn all these great men and women for any kind of ethical failing. Many of them were exemplary figures with a deep love of European culture; they lived in truly perilous times and acquitted themselves admirably. I have only sought to demonstrate what they lost in the process, which was any hope of maintaining control over their scientific research agenda. What began with Haber in the First World War was brought to its close by the end of the Second."

Judging this to be the right moment, she turned to Hera. "Speak up, sister. You and I have rehearsed this part of the story many times. Tell Abdullah what you and I concluded from our reflections on these events."

Hera still looked exhausted and preoccupied with other matters entirely, and she seemed reluctant at first to intervene, but I knew she couldn't resist in the end. "All right, then. Let's try to imagine a counter-factual case. Suppose that the discovery of nuclear fission had occurred at a time of less drastic social instability, in a set of circumstances that offered those scientists more options in life than the only one available, namely, running for their lives from a gang of savage murderers. As you said earlier, Gaia, as soon as they learned the results of the Meitner–Hahn experiment, many of them knew that science had arrived at a fateful turning point in its development.

"I honestly believe that under different circumstances they might have sought to stop the existing process of discovery in its tracks, indefinitely, by refusing to perform or publish the results of any additional experiments. Remember, this was a community that was relatively small, whose members knew each other intimately. We do know that Szilárd tried, and failed, to do exactly such a thing in the 1930s, especially in relation to the work being carried out in Paris by Frédéric and Irène Joliot-Curie. Because, of course, there was an obvious risk that those scientists and their expertise would fall into German hands one day soon. Who knows? It's not entirely impossible to imagine they might have succeeded under other circumstances."

As I watched, the color drained from al-Dini's face. "I appreciate your frankness, Hera," he remarked. "And I'm most grateful to you for telling me so clearly that you and Gaia and your associates are engaged in a campaign right now to bring the process of scientific discovery to a halt. At least, that is what your comments strongly imply. What I remain confused about is why you have taken the trouble to haul me here, all the way from the Mediterranean. Had I known that you were intending to prevent me from doing further work, I would never have placed myself in your hands. Was it your purpose to kidnap me, in effect, in order to take me out of circulation as a working scientist?" The bitterness in his tone was all too obvious.

I expected either or both of the sisters to lash out at him, but neither did. It was Hera's weariness, perhaps, that was most evident in her measured response. Gaia was obviously ready to substitute, and for a moment, as I glanced in her direction, I opined that she might offer al-Dini and his family seats on the next plane headed east. But Hera waved her off with a slight movement of her hand.

"No, there is no such implication in what I said. I tend to use precise expressions, and the nuances can get lost in oral exchanges. Clearly what I said upset you, and for that I apologize. Recall that I was envisioning decisions that this group of distinguished scientists itself, of their own free will, might have made. I have no coercive scenario in mind. I only ask you to imagine that this group might have agreed among themselves to suspend its work until they had figured out a way to resume it in secret, securely hidden away from the malevolent state's prying eyes and the corporation's eagerness to exploit its downstream products for profit."

"I, too, must apologize, Hera, for hastily jumping to conclusions," al-Dini immediately added. "That is indeed a much different prospect than the one that leapt into my own mind as you spoke. I still don't fathom why you might believe there was the remotest chance of their succeeding at such a quixotic endeavor. Nevertheless, as an idea for debate, it is unobjectionable."

"Thank you. Obviously, Franck, Szilárd, and their supporters within the scientific community didn't succeed in influencing events in 1945. Ten years later, a few of them, led by dear old Einstein, tried again: He signed the 'Russell–Einstein Manifesto' two weeks before his death in 1955. The hydrogen bomb had been successfully demonstrated only a few years before. The manifesto's signatories protested against the possibility of the use of what they referred to as 'weapons of mass destruction'—a phrase widely used and abused a half-century later—in a future war. But the cold war was in full swing at that time, and the group's efforts, as well as the entreaties from the later Pugwash Conferences, fell on deaf ears.

Noble, futile gestures, made without any real hope of realization, I'm pretty sure."

"I agree, both noble and futile. As I recall, the Cuban missile crisis occurred not many years later, in 1962, I believe, when the world narrowly averted what that manifesto's signatories had anticipated."

"We appear to have made some progress at last in our little colloquy," Gaia noted. "We seem to be in agreement that if leading scientists were to propose to put a moratorium on their work, out of fear of the consequences arising from its application for terribly destructive purposes, they might be morally justified in doing so. In 1945 a large number of them argued that a technical device they themselves had created should never be used. I can tell you, Abdullah, that Hera and I have drawn much inspiration from their example."

For the first time in the whole of these two sessions al-Dini actually laughed, heartily. "As they say, the penny has finally dropped for me. You must think me dreadfully obtuse. I just now realized that we have arrived back at where we started last time. This is the basis for your rule against disseminating at least some new scientific results to the wider community of scientists and to the public."

"Let me give you an analogy. At the time, about a half-century ago, when large corporations ruled a good deal of the world, a favorite practice of financiers was to take a publicly traded corporation and return it to private capitalist control, thus avoiding the most onerous provisions of the securities regulations. We have taken our cue from this impressive precedent. We are in the process of re-privatizing the scientific enterprise. I say 're-privatizing' quite deliberately, because what we are doing is returning modern science to the conditions where it first arose—in the makeshift private laboratories of its fiercely dedicated seventeenth- and eighteenth-century aficionados. With the slight modification that our labs are not makeshift in any sense, but rather state-of-the-art, as you've seen for yourself."

There was no immediate comment from al-Dini. Undoubtedly, he had realized that the conversation in which he had been engaged had taken both a serious, and a personally relevant, turn. Finally, he shifted in his chair and spoke in barely audible tones. "And what if, tomorrow, say, you decide—in your role as the owners of this private enterprise—that you are bored with it, or unwilling to carry on for whatever reason. Will you give us all a severance package and show us the door? Or what if a few of us happen to displease you, for whatever reason? Will you put us and our families on a plane and send us back to where you found us?"

"Slow down a bit, please," Hera responded quickly. "If these possibilities really concern you, we'll talk about them. We have built safeguards into the constitution of Solomon's House that should allay any fears you may have along these lines. Gaia, I want you to set up a meeting, as soon as possible, for you, Dr. al-Dini, and the director of the faculty association at our university. I want you to review the rules and procedures with him, including the protections regarding the academic freedom of our scientists."

"Sister, your wish is my command. Consider it done."

Hera turned to face al-Dini. "This is a matter of sacred honor for us. But I detect in your remarks a lack of understanding as to what is so important about what some may regard as the quixotic venture on which my sisters and I are embarked. So I want to plead with you to agree to return for one more cycle of discussion. Will you give me your word that you will respond favorably to the invitation?"

Hera can be persuasive when she wants to be. The earnestness of her entreaty must have been obvious to al-Dini. Perhaps his thoughts also cycled quickly back to the time when he and his family were in distress, and to the rescue operation we engineered. Or to the warm welcome he had received from the other distinguished scientists in his section after his arrival here, and the generosity they had displayed in equipping his labs with everything he could possibly require for his work.

"I will be most happy to accept such an invitation."

At which point Gaia called the meeting to a close and we all went about our business.

Part Three

Mammalian Mothers

The evolution of mother love was essential for the evolution of intelligence.

C. A. Pederson (2004)

I had never seen Hera so frightened. "Go after them! Bring them all back! I admit they have a right to be outraged at what happened to Lucetta, but fleeing from Yucca will only make things worse. No matter where they're headed they won't be safe. Surely they realize that?"

She was addressing a small group of us who had hastily assembled in our command center an hour ago in response to the emergency alarms, which had sounded when the perimeter systems automatically reset themselves after being disabled sometime earlier. The personal tracking system in our computers quickly told us what had happened: Two of the young women of our Second Generation, Lateefah and Myra, had fled from Yucca Settlement during the previous night, along with their two young infants—and Io.

I replied, "Listen. We're assembling a pursuit unit now. We'll have to travel by truck caravan; the weather's far too foul to put a plane or helicopter aloft. But we can't follow them until we know their direction of flight."

I had placed a call to Lucetta ten minutes ago, and now I glanced in her direction as she and Ming entered the room. Along with a few other young men and women of their cohort, they had just finished an initial round of combat training. I got the impression that they were eager to apply what they had learned.

"Sometimes Lateefah is too smart for her own good," I said to Hera. "She expertly disabled the perimeter surveillance and personal tracking systems on our main computer system before crossing through the gates. She also took one of the vans from the motor pool that happens to be equipped with our most advanced stealth technology—there's absolutely no trace of the vehicle on our GPS screens. We won't pick up a signal we can track until one of the passengers exits the vehicle, and that hasn't happened yet."

"Damn her!"

I don't think I had ever in my life heard Hera curse a member of her tribe before.

"I *must* be able to talk to her, Lucetta, to persuade her to come back. Many of the women in her generation look to her as a leader. What if she's left directions behind and some of them decide to follow her? It's a catastrophe." She stopped and her voice fell. "Thank you for agreeing to go with the team."

At virtually the same moment, a staff member, coming on the run from the motor pool building, entered the command center. "We've just found a note addressed to you, Hera. It seems to have been dropped behind a barrel near the vehicle entrance—or maybe it was blown there by a gust of air when the van exited."

She seized it from his hand and opened it. "It's from Io. The writing is so bad I can barely make out the words; it looks like it was scribbled as she was walking. Here's what it says: 'Myra and Lateefah asked me to go with them. I couldn't alert you. I'm going to try to bring them back. Io.' Here, read it for yourself."

I scanned it and glanced back at her. "Do you believe what it says?"

Her gaze was impassive. "Yes. Yes, I do, unequivocally."

Suddenly the senior officer on duty looked up from his monitor. "We have a signal!" He paused, interpreting the information being received. "We have three signals from the implanted personal locator devices—presumably, Io's, Lateefah's, and Myra's. At least, I ...."

Hera interrupted. "Then they must have stopped and gotten out of the van. Where are they?"

He pointed to the nearest screen. "They're at the southern edge of what was Joshua Tree National Park. The location is about two hundred miles almost directly south of us on a straight-line basis. But, of course, the roads are not direct."

I turned to Hera and said, "I know where it is. They probably drove through Las Vegas and then kept on going straight south on Highway 95, then swung west on Interstate 10, since those are by far the best roads."

Hera was grim. "I think they're probably heading for Jalama Beach. But something must have happened to cause them to stop and get out of their vehicle."

Lucetta turned to the officer. "Put the latest area surveillance maps up on another screen. What do we know about who's in the vicinity?"

"It's already done," he replied. "Step over here."

The four of us examined the ground photos for a moment and then the words just popped out of my mouth. "Oh shit!"

"What is it?" Hera exclaimed.

I touched her shoulder. "Take it easy. There's been a large religious colony camped at two separate locations in the old national park there for the past six months. We can practically see inside their tents with these satellite cameras. We only get one scan of the area every twenty-four hours, but as you said, we can surmise that our group's vehicle is no longer moving and that the adults, at least, are somewhere outside it, since otherwise we wouldn't have been able to pick up their signals. I'm afraid it may not have been a voluntary stop."

As I stared at the images for a few moments longer, Lucetta said, "I'll get the team assembled and we'll leave immediately."

Hera grabbed her arm. "Lucetta, be careful. I mean, in approaching them, if they're being held captive or something. If they are prisoners, and if you charge in full bore, they may be hurt or killed."

"Yes, of course, you're right. We're all just guessing now, but I'm assuming that they wouldn't halt their journey in that odd place unless they were being detained by some people in the religious colony. I promise; we'll be careful in approaching them if it does turn out that they're being held captive."

She hugged both of us. "Go quickly then. Bring them back to me. I'll be praying for you."

As the group of us ran in the direction of the motor pool building, where the final preparations for our pursuit were being made, I mentioned a new idea to them—inspired by the knowledge that we were off to rescue my mother and by my familiarity with some bizarre episodes in her life. The rest of them looked at each other, smiling and nodding in agreement, whereupon Lucetta radioed to the warehouse, asking for some supplies to be assembled in a hurry. When I had first imagined going after the group, before we knew in which direction they were headed, I had immediately considered the possibility of taking some horses with us in case we wound up having to give chase over rough terrain, and I had placed a call to our Hutterite stablemaster. Now I asked Lucetta to call him again. "We've decided we want to take four of the heavy horses instead. Is that OK?" She looked at me and nodded as the answer came back.

By early afternoon we were on the road with our caravan, consisting of a powerful tractor-trailer unit ferrying the horses, and two armored passenger vans carrying a fifteen-member combat team made up of a mix of our own young people and some seasoned US Army veterans from the Beatty staff. One of the two vans had left first so that its occupants could act as advance scouts for the rest of us. We were all armed to the teeth, although somehow I had the feeling that guns would not figure prominently in our rescue mission. I had given a few of the others in my van some simple tasks to perform—involving a bit of needlework—while our vehicle sped along the highway. As they sewed, I introduced them to the novel plan of attack.

I sat next to Lucetta as she drove, and occupied myself by poring over my terrain maps and historical atlas so as to refresh my memory about the specific landscape we would encounter. The area once known as Joshua Tree National Park lies about one hundred and twenty-five miles directly east of Los Angeles. Ironically, it was appropriate that we were headed there to, most likely, confront the members of an evangelical Christian religious sect because it was some of their forebears, the nineteenth-century Mormon migrants, who had given one of the area's distinctive plants its common name. The plant they baptized with the odd name "Joshua Tree" is in truth *Yucca brevifolia*, a tree-sized member of the lily family. Legend has it that the Mormons interpreted their own westward journey with the help of the Biblical Book of Joshua, where Joshua is described as leading the Israelites to the promised land of Canaan. To the weary Mormon travelers, these great yuccas, with their large branching arms, that they came upon in southern Colorado seemed to be welcoming sentinels, beckoning them onward to their final destination.

The general area of the Park is of interest because it lies at the intersection of two ecosystems—the Colorado Desert in the eastern half and the Mojave Desert to the west; the latter is higher, cooler, and a bit more moist.

Humans have lived in the area for about ten thousand years, sustained by both small and large game and abundant plant resources, including acorns, the pods of the mesquite bush (a member of the pea family), pinyon nuts, berries, and cactus fruits. *Yucca brevifolia* itself provided its tough leaves as materials for baskets and sandals, and nutritious seeds from its flowers. By the time the white miners and settlers showed up in the nineteenth century, the native tribes included the Chemehuevi, the Serrano, and the Cahuilla.

An hour into our journey Hera was on the radio. "The signals show that they're moving, but very slowly! They're heading north into the park area."

I radioed back, "Keep us posted. I especially need to know their rate of speed and whether or not they leave the old park boundary at the northern perimeter."

Five hours later her voice came through again. "They've halted again, still definitely in the southern portion of the park, still well within the old park boundaries."

I cheered up instantly. "I've hiked in that area, Hera. I'll bet they've ended up at the place known as Cottonwood Spring Oasis." At various times over the years, that spring had provided copious supplies of water to native tribes and the later immigrants who stopped there. The oasis itself is a collection of the indigenous desert fan palms and some cottonwood trees. "It's only about six miles from the old southern entrance to the park, and the road into the park should still be passable for our vehicles. I'm betting our party will be kept there for a while. I'm sure that at least some of the religious group, maybe even the leadership, has its base at the spring. It's just too good a water source not to figure in their desert holiday."

A short time later I radioed Yucca Mountain again. "The advance team just contacted us. They've reached the southern perimeter of the park, just at the turnoff from

Interstate 10 to Joshua Tree, where they found their vehicle, burned to a shell—but there's no sign of bodies. Our advance team is going to go through the southern entrance and then see if they can get close enough to scout the Cottonwood Spring area under cover of darkness. Our group and the horses won't arrive there for another couple of hours, when it will be well after nightfall. We'll rendezvous with the advance team when we arrive and make a plan based on their report."

About midnight we reached the southern entrance to the park. Keeping in radio contact with the advance team, who had told us they had the camp under surveillance, we proceeded toward Cottonwood Spring Oasis, moving slowly on account of the poor condition of the road and the howling windstorm that nicely masked the sound of our engines. An hour later we stopped the vehicles near the ruins of the old visitor center, about a mile from our destination, where Ming was awaiting us. Lucetta and I leapt from the cab of our van and ran to him.

"What have you learned?"

"Our team is in place around the periphery of the encampment. It looks like there are about one hundred and fifty in all, living in tents, plus some mule and horse teams with a fair number of wagons and a few dogs. We've followed orders and kept our distance, staying downwind of the dogs. We're still receiving some signals from the locator beacons, so we think we know where our people are being held. We believe they're in a cave dug into the hillside above the oasis; we have a clear line of sight from the surrounding hills with our night vision gear and can see that the entrance to the cave is guarded. But we can't get close enough to tell what's going on at the campground itself."

"Do we have to be concerned about the second group of people who are camped somewhere to the north of here?" Lucetta asked.

"No. We've had time to send a small squad up there to have a look, and there's no indication of any movement back and forth between the two groups. By the way, in case you're interested, there's still an old road sign near the second camp. The place is called Fried Liver Wash." We all guffawed at the image this conjured up in our minds, which relieved the tension a bit.

Ming continued his report. "I have to say we're becoming quite worried about Io: Her beacon stopped transmitting just before we arrived. The team thinks we should launch an operation at the cave right away."

"Thanks, good work," Lucetta replied. "I'm also worried about Io's signal failing, but I'm not willing to risk an attack just yet. Her beacon could have stopped for any number of reasons. And we promised Hera we'd proceed with caution."

I turned to the rest of our people from the second van, now assembled nearby. "Unload the horses. Give them some food and water. Get the gear from the vehicle."

While the horses were being refreshed, Lucetta and Ming reviewed the placement of the commando team around the oasis, who were under strict instructions to hold their positions until other orders were given. Weapons were checked and readied. I made a final call back to Yucca Mountain and told Hera that we were close to the camp—and not to expect to hear from us again until after dawn. Then we led our mounts by foot toward the oasis.

The horses assembled by the stablemaster for our mission were Percherons. The one assigned to me was an immense coal-black stallion, standing nineteen hands at the shoulder and weighing almost two thousand and seven hundred pounds, aptly named Hannibal. The others were a white mare and two dappled-gray mares, all somewhat smaller than the stallion but still of great size. After walking a short distance, we found what I had told the others to look for—the entrance to the wash that begins about one thousand feet from the oasis, then angles away from the spring through a ravine before taking an abrupt turn to the north and leading directly back toward it.

The desert washes in this park are wide paths of fragmented sandstone, made up of tiny pebbles leached by rain and wind from the surrounding slopes that yield easily to one's footfall. Our beasts snorted with pleasure as their huge hooves first touched the sandstone path that bore their great weight so gently, and it was with some difficulty that we prevented them from breaking into a spontaneous trot. Fortunately, the soft surface underfoot also muffled the sound of our horses' steps so completely that our little band moved along in near silence.

Lucetta called a halt when we reached the turn in the wash. It was now about a half-hour before daybreak, and the storm had finally abated; peering around the turn, we could see the glowing embers of the campfires. She did a final radio check with the team that was in place on the surrounding hillsides, then distributed the crude costumes that had been sewn together during the long ride from Yucca Mountain. Still concealed behind the low valley wall, we outfitted our horses and ourselves, then slowly walked the beasts around the turn and mounted them, using a boulder sitting at the edge of the wash to reach the oversized saddles. I and the other three riders sat astride them in a row and gazed ahead, our heads almost a dozen feet off the ground. Lucetta and another three stood in a line in front of us, one in front of each horse, draped head to foot in black gowns, their weapons held at the ready beneath their cloaks. We were now in sight of the encampment, although as yet no one stirred there.

As the first light of dawn broke over the scene before us, I looked to my right and left at my companions. I wore a long black cloak that flowed down to my ankles, with a hood that covered my head so that my face was entirely concealed. Beside me was the white horse, its rider draped in a similar gown of pure white. On the outside ranks were the two dappled grays—one, horse and rider alike, was covered all in red, and the other in pale green. I was no expert in Biblical lore, to be sure, but I was fairly confident that the residents of the nearby campground would recognize the Four Horsemen of the Apocalypse when they sighted us.

I felt my mount's body stiffen and saw his ears rotate forward as the first figure emerged in the still-dim morning light, carrying pots for water to be filled at the spring. My horse gave a low rumble in his throat, and I stroked his nape, speaking softly to him. The light grew quickly now; standing motionless in the wash, we knew we must be clearly visible to them.

The figure suddenly looked up in our direction; she froze, dropped her pails of water, and sank slowly to her knees, emitting a long, low moan. Her hands came up to cover her face. Another emerged from a tent and cried out, waking the remaining sleepers. As they staggered forward, each copied the first, falling to their knees, hands clasped in supplication, facing the monstrous apparitions standing silently a short distance away.

When their movements had stopped, at a signal from Lucetta I pressed gently on my mount's flanks with my knees and gave my command: "Hannibal, walk on." The great beast snorted and slowly ambled forward, followed by the others. When Lucetta signaled again with upraised hand, we called a halt by pulling on the reins, stopping the horses a few yards from the assembled group. I saw now that they were outfitted only in filthy gowns or long shirts, their hair matted and unwashed. Hannibal snorted again and their bodies recoiled at the sound.

I had been designated to deliver from my high perch the lines we had all scripted together. I waited a full thirty seconds before speaking, my words finally emanating from somewhere behind the hood that concealed my entire head and face.

"Who is your leader? I command you to come forth."

An older man, large-boned and stocky, with long tousled white hair and a full beard, rose awkwardly, his knees wobbling uncontrollably. He moved forward, and once again sank to the earth at the head of the group. He was shaking violently from head to toe.

"Know ye who stand before you?"

The man sought to reply, for his throat and lips moved, but no sounds came forth from his mouth.

I waited.

Finally he stammered, "You are the four whose coming was foretold in the Book of Revelation. But what do you want of us?"

"That you shall soon learn. Do not question me, but only answer. What have you done with the others who are not part of your group?"

"They are in a cave, up on the hillside behind us, sire. All but one of them."

"Hear me. Name three of your followers to lead my demons to them." Here I pointed to the members of our team standing next to the horses, their hooded faces invisible. "Wait. First answer me this. Have they been harmed?"

The man spoke more easily now. "No, sire. Those kept in the cave under guard are as we found them yesterday, to the south of here, near the highway. We are awaiting a sign from our Lord, in answer to our prayers, to tell us what we must do with them."

I leaned forward to peer down at him from on high. "Then have your people go now and release the captives. And if I find that so much as a hair on the head of any one of them has been touched, I will open a crevice in the earth before you and cast you into the pit of everlasting fire below."

The three campers nominated by their leader got up and walked slowly toward the hill, followed by three of our four demons. As Lucetta passed the campfire, she inserted a small metal implement into the coals, and left it there.

I addressed the leader again. "You spoke of all but one of the captives. Where is that one?"

"She perished last night at the stake, sire, after confessing to being a vile and blasphemous witch, who had flung curses at us in many tongues that we did not understand. She boasted of her service to Satan. Look; here I have the proof of her wickedness, sire. Look at this remnant of her body that survived the consuming fire."

He held out his hand and I instantly recognized the remains of what appeared to be one of our locator beacons. I figured that the blackened device must be the one formerly implanted in Io's body.

"Drop it. Where are the ashes from the fire in which you burned her?"

"Over there, sire, at the edge of the palm grove."

Motioning to the other riders to hold their position, and gesturing to the fourth demon to follow me, I directed my stallion to the place and stopped before the ash heap. Nearby I noticed the piles of dried mesquite branches. That wood burns furiously and white-hot; perhaps her death had been mercifully quick. Using an empty canvas water bag lying a few feet away, my companion carefully swept the ashes inside it.

I dismounted and walked my horse back to my former place, standing directly in front of the still-kneeling leader, who began to tremble so violently that his teeth chattered. Then I turned to my black-clad companion. "Fetch me the iron from the fire."

As I waited I thought back to my quick visit to my suite at Yucca Mountain while the team was making final preparations for the dash to Joshua Tree. I was looking for something confiscated long ago from Io's belongings during our Caribbean days, at the time when Hera had her sister transported to Salt Cay Island. It was a branding iron, made of brass, with a simple design at its tip: 666.

I took the glowing iron and stepped in front of the leader; at a sign from me, my companion tore the shirt from the man's body, and immediately I plunged the red-hot metal against his chest. The air stank with roasted flesh and hair as he uttered a piteous scream. I bent down to the writhing figure, lifting his face to mine until the two of us were separated by mere inches.

The words hissed from my mouth, still hidden by my cloak: "She … was … my … *mother*!"

§ § §

Meanwhile, following their guides, the three rescuers had scrambled up the hillside and halted, directing the pilgrims to approach the shallow cave and summon away the small group of guards posted there. One of our people radioed Ming's team: "Move in, quickly." Having ascertained that no other guards remained inside, Lucetta strode to the aperture and called. A few moments later, Lateefah and Myra emerged, haggard and wan, carrying their babies, and two of the rescuers surged forward to meet them while Lucetta stood guard over the group of pilgrims with weapon ready, greeting Ming's team as they came on the run. Then the same words issued from Lateefah and Myra simultaneously: "Where's Io?"

Lucetta called me. "We have the mothers and children and they appear to be fine. Io's not here. Have you found her?"

After a brief pause I answered, my voice choked and barely audible. "She's dead. They burned her at the stake last night."

The rescuers took the infants from their mothers, who then collapsed into each other's arms, sobbing. Lucetta spoke to the team. "Take the babies back to the vans and give them a medical checkup. The rest of you, search the cave and tell those people they can rejoin the others down at the campsite." To the rescued prisoners she said, none too kindly, "You two come with me," and she led the way back down to the camp with the rest of our pursuit team.

When they reached the base of the cliff again, they found the terrified colony of pilgrims huddled on the ground, shielding their faces from the scene in front of them, as I raged through the campground on my immense stallion. Before remounting Hannibal, I had reheated my branding iron in the fire, and now I plunged it against the trunks of the encircling California fan palms, one by one, marking them with the feared sign of the devil.

"My iron harms not these trees, which love fire, but you shall tremble at the sight of my signature," I screamed at them. "You shall remember this day and moan at the thought of it even on your deathbeds, for you shall know that I await you below amidst my welcoming torments."

I brought my mount to a halt in front of the leader. "Now go from this place and never return! Go! Find your fellows and leave! This grove shall be sacred to me and my demons forever. Should I ever find you within a hundred miles of it hereafter I shall strike you down where you stand!"

As the pilgrims scrambled for their wagons, herded by our team, and carrying their groaning leader to the head of the column, I saw Lucetta and her troop walking over to the remaining three riders, who dismounted and greeted the rescued prisoners. Finally I joined them, my hood now thrown back; the anger and grief that consumed me must have been visible on my face. I looked at Myra and Lateefah. "What happened?" I snapped at them.

Still struggling to regain her composure, Lateefah spoke. "She saved our lives, Marco. I'm so, so sorry. Will you ever be able to forgive us?" I glared at her without responding, and she continued. "Shortly after we were brought to this camp, we were led to the cave, with guards posted outside. A few hours later, the leader returned and told us that he had prayed for guidance, and that the Lord had spoken to him, telling him we were emissaries of Satan, come to snatch away their immortal souls. He said that the three of us would be burned at the stake and that the babies, who were blameless, would be adopted by the group and raised as Christians."

"So how did you two manage to avoid the sentence of doom?" I asked, not trying to disguise the bitterness in my voice.

"As soon as the leader had finished, Io lunged at him, screaming at the top of her voice, scratching his face and tearing at his clothes. His companions grabbed her, but she continued to yell at them, fighting them, until they bound her with ropes. She was incoherent, uttering words I didn't understand—they were a mixture of Hindi and Balinese, I think, judging from the sounds of them. When they covered her mouth at one point, the leader told his companions that she was 'babbling in tongues.' Then suddenly, free to speak again, she switched to English, telling them that she was sent from Satan to destroy their faith, that we two were under her spell, that she had snatched us and our babies from a church, where she had come upon us praying, that she was preparing to sacrifice us in a demonic ritual, asking them if they had an altar on which she could carry out her plan.

"At that point they dragged her out of the cave, the leader telling his companions that she must be interrogated further in front of the whole assembly. We never saw her again. They never came back for us." She halted, sobbing, and collapsed on the ground.

Lucetta turned to Myra, who had stood mutely nearby, her eyes downcast. "She knew we would be in pursuit. She was buying some time for you—barely enough time, as things turned out. Just what did you two think you were doing, and where exactly were you going?"

Myra continued to stare at the ground in front of her. "We were so angry at what had happened to you. We had started to hear all kinds of excuses for what Rainer and Kenji had done, from others in their circle, about 'female oppression' and other crap. About how those two were just trying to 'make a point' and really didn't intend to harm you, just scare you a bit. Anyway, we decided we had to make a point, too."

"Hera guessed that you were headed for Jalama Valley. Was she right?"

"Yes, of course. We figured we'd be safe there, and we thought we could make it easily. We knew that Interstate 15 going west from Las Vegas had been sealed off as part of our isolation plan, so we headed straight south on Highway 95 until we got to Interstate 10, which is still quite well maintained since it's the only main east–west road left in the southwest. We ran into a gang that had set up a roadblock just past the exit to Joshua Tree; they shot at us with rifles, and we managed to double back to the exit, heading north for a few miles until our punctured tires gave out.

"Almost as soon as we got out of the vehicle to assess the damage, we were set upon by another large contingent, who appeared suddenly over a rise and surrounded us. We thought of putting up a fight, but we knew that if we did the babies might be harmed. So we surrendered and were promptly deposited in one of their wagons. They then set fire to our vehicle and forced us to accompany them to this campsite. You know the rest."

As she finished, the last of the pilgrims were straggling out of the oasis. I said out loud, to no one in particular, "I shall return here as soon as I can to better secure the site and set up a memorial. I'll talk to Hera first about what kind of shrine would be suitable. Which reminds me, I'd better call her right away. But all I'm going to say is that the rescue operation was a success, and that she'll have to wait for the details until we return. I want Lateefah and Myra to be with me when I tell her what happened to Io."

§ § §

Shortly after we returned to Yucca Mountain, Athena told me that Hera had sequestered herself in her suite and cut off all outside communications, except for regular visits by her sisters. As I would discover, she had been occupied with them on a single project—the design and construction of Io's memorial.

About three months later Hera sent word that I was to meet her on the mountaintop. Alighting from the car of our funicular railway, where I had been jammed in among diverse construction materials, I walked past the helicopter landing pads, through the dense forest of satellite uplinks, beneath the whirling radar dishes and remotely controlled weapons batteries, to the summit's southern end, which overlooks the multicolored Calico Hills. As I neared the point, passing through a break in a newly erected granite wall, I halted suddenly, astonished, standing dumbfounded until she approached.

I was facing a huge hemisphere, looking like one-half of a massive, transparent egg, sitting on its middle. Inside stood a large telescope, wrapped tightly in a protective tarpaulin. I had to shout over the cacophony of pneumatic drills, forklift trucks, and dozens of workers. "None of this was here a few months ago."

"The crews have been here day and night. The basic structures are in place. But we'll do the fine finishing steps at a much more leisurely pace, with loving attention to detail. Let me show you around." Passing a sign carved into a granite panel saying simply 'Io's Walk,' we entered a semi-circular pathway that wound around the mountain's edge.

As we navigated the path, admiring the dome's simple beauty, she remarked, "By the way, I heard about how you drove the pilgrims from their campsite at Cottonwood Spring, forbidding them from ever setting foot on the spot again. The others told me what you said during the return journey about creating a memorial for Io there. I want you to do that, Marco; I want you to build it as a personal tribute to your mother. You must do your own design. When you've finished, I will pay a visit to the site with you."

She stopped and moved toward me, tears streaming down her cheeks, and we held each other. Then she turned to face the dome. "We always intended to put our observatory here. Now the entire facility will be dedicated to Io's memory. When the telescope isn't being used, it will have a default position, trained on Jupiter and its moons. What better tribute could there be for someone whose spirit soared so high and so far? Let's go inside."

Inside the dome we entered a soundproofed control room, facing a bank of computers and monitors covering one entire wall. Adjacent to it was a small amphitheater, where we took two of the seats. As I glanced around I saw mounted on one wall in an elegant frame the original of Io's scribbled note, left for Hera on that fateful day when Myra and Lateefah and their two children had fled from Yucca Mountain. She saw me notice it.

"The saddest part of this whole wretched business is how contented Io had been since she joined us here. What a cruel turn of events."

"Not only that," I replied. "She spent every day in the midst of our young tribe, and they responded to her with great affection. Some part of her fiery spirit broke through the fog of her medication and poured over them. She was a great performer of many roles, and sometimes they just flocked around her, delighting in her play-acting. During the last few months I have witnessed their collective grief first-hand, many times."

"Do you remember, Marco, long ago, when we were living in the Caribbean, I related the episode that occurred during one of the raucous private parties we sisters were so fond of? The one in which Io wrote and starred in a little skit about salvation, where an overweening supplicant stands before St. Peter and demands entrance to Heaven? She took both roles, of course. The weary guardian of the Pearly Gates sends a DNA sample from her soul to the Holy Laboratory, which detects numerous genetic modifications in the product, so Peter refuses her application and banishes her to the transit station where those consigned to Hell are being assembled. There Satan—again played by Io, after a quick costume switch—tells her to take a number and wait, because he is utterly swamped by the hordes sent his way by God and destined for the devil's fiery kingdom far below."

"I recall well your retelling of it, and the sheer playfulness of the performance she staged for her sisters helps to balance in my mind the other and more ominous career she took up once she landed in Bali after her escape from Salt Cay Island. The innocent fun had vanished by the time she assumed the role of Durga, the Hindu goddess of death: The slaughter she staged in her encampment was all too real. But I have to say she was eminently believable in that other role, too."

She didn't—or couldn't—reply, but instead just sat there, sobbing, until she summoned the will to speak again. "It's past time for us to have a serious talk on another matter, but first I want to play a piece of music for you." She saw the look of amusement on my face and smiled. "You know me well enough by now to have anticipated that there would be a music system somewhere nearby. I'm glad I didn't disappoint you."

She pushed some buttons on the remote and then paused. "I'd been dreading for a long time the day that I knew had to come, the day when I lost the first of my sisters. Before the girls brought the news to me about what had happened to Io, I had been delirious with joy at the success of your rescue mission; that made the subsequent blow so much harder to bear, of course. In such times I always turn to music. This is something from a piece you know well—Giuseppe Verdi's *Requiem Mass*—but not the passage that struck you so, as I recall, when you were watching the surveillance videotapes of Io's bizarre ceremonies in Bali."

"That was the *Dies Irae,*" I replied, "which seemed to fit perfectly the scene I was then looking at."

She nodded and said, "It's extraordinary, even for an unbeliever like me, how perfectly Verdi seems to match his music with the text of the Catholic mass for the dead in this work. Of course, I respond to the music's emotional force; indeed, it seems to me, as this work progresses, that his music has shattered the religious context in which this type of work had always been framed until then. A mere thirty years later, Mahler's *Song of the Earth,* that anti-requiem, completes the process.

"Anyway, this section is very, very different from the *Dies Irae.* It's the second part of the *Offertorio.* In the Latin text that is being sung, God is asked to bring the dead back to life. Since you carried Io's ashes home to me, I have played these passages over and over, despite the fact that as I press the button to start the recording, I know I will be unable to stop myself from beginning to weep uncontrollably. Listen."

She had selected one of our favorite renditions, as I knew she would, and as it began, the unaccompanied tenor voice of Jussi Björling was heard:

Lord, in praise we offer you
sacrifices and prayers,
accept them on behalf of those
whom we remember this day:
Lord, make them
pass from death to life,
as once you promised to Abraham
and to his seed.

She played it twice, and then she began to speak softly, almost inaudibly, not looking at me, as if in a reverie. "Verdi creates for us in these notes a feeling of such desperate and unbearable longing that it seems we, too, surely must perish there and then. As I listen, my mind is at first seduced by the music's emotional force, pulled uncontrollably toward the abyss, toward that truly abominable lie, the one that goes, 'you may let go now, for you shall be reunited after death.' It holds me in its unbreakable grip, I am crushed by sorrow, I literally gasp for my next breath, I want to be dissolved into the chorus and cry out with them, '*salva me, fons pietatis,* save me, font of pity!'

"Then the miracle occurs. His music has thrown me down so brutally, down against the very foundations of my being, that I bounce back, seemingly against my will. My sorrow remains undiminished, but now it has a different object. Through my grief I understand that these years my sister Io and I shared here on earth is all the time we will ever have together. There will be no second chance, no facsimile of life, no ghostly post-mortem venue where all is forgiven. Through it I am compelled to face the truth that by my acts, I added to the great burdens she carried as a result of her illness, and that if I had acted differently she might have been spared a small measure of the suffering she endured. These failings will remain with me always, unforgiven. There is no hope for redemption in some other pale and insubstantial dimension for the wrongs we do in this one. That is the real nature of the sorrow I bear upon her death."

Her grief was contagious and I sat there in silence, waiting to be released from it.

"I need to say something else to you," she continued, "and it fits the occasion well. There's only ever been one good response to the death of those who have been close to us, and that's the joy of seeing new life spring up upon the earth. Lucetta came to see me yesterday. She wants to have a child with you, and she asked for my blessing. I told her she didn't need it; she said she knew that, and then she reiterated her request. I gave her what she sought, most happily."

In mid-summer 2066, Hera convened a meeting of the Sujana Foundation's board of directors, consisting of the eleven surviving sisters and me, in a lovely room overlooking the Pacific Ocean at our summer retreat, Jalama Beach, in central California. The Foundation legally controls all the physical assets of the Yucca Settlement, an investment portfolio amounting to tens of billions of euros, and leases from the US government through which we manage various properties, including what's left of the city of Las Vegas. The one thousand children of our Second Generation had recently celebrated their twentieth birthdays. Since some of the young women in that cohort had already given birth, we were starting to be fruitful and multiply. The sisters could no longer avoid the question of political authority.

Hera asked her sister Hecate to table an opening proposition to generate debate, and assigned Rhea the job of attacking it by playing the role of devil's advocate.

"My opening gambit is a simple one," Hecate began. "Collectively, the members of all three generations of our kind make up a sort of tribal entity. We sisters are, in a legal sense, the guardians of the individuals who make up what we call the Second Generation, as recorded in agreements their biological parents executed when they donated their eggs and sperm to our Foundation. Thus we represent a kind of council of elders for the tribe, and we rightfully exercise control over its collective assets because it was we who accumulated those assets. For the same reasons, the members of the council control the security forces that protect our resources. The terms of our various leases—which I manage on your behalf—stipulate that our government partner will not challenge our exercise of authority within the boundaries of our territories. The sole limitation is that we cannot impose and carry out a judicial sentence of death.

"All of the other persons who aren't tribal members but who reside within our borders, including the Hutterite colonies and the technical staff based in Beatty, as well as the personnel of Sujana University in Las Vegas, and their families, have voluntarily acknowledged our authority. All of those groups have formal structures for discussing matters of concern among themselves, and our management team seeks to negotiate solutions to perceived problems. When we cannot come to terms with demands from certain individuals, we invite them to leave our territories. Some do, but as you know, we always have more requests from outsiders who want to be admitted than we can accommodate.

"This is our political structure, if you will, and it seems to have worked well, on the whole. My recommendation is that we just carry on, although I also think we should set up a committee of the board to prepare a transition document that examines future alternatives. I think it'll take us at least five years to choose a preferred option."

Artemis interrupted just as all others looked to Rhea for her riposte. "You've forgotten, sister, that we've been experimenting with another political structure for two years already. We do have, as you well know, an Assembly of the Second Generation, where matters of governance are presented and discussed—and sometimes rather hotly debated. Our council of elders receives proposals from the

Assembly, and as far as I can recall, we've never refused to ratify any of them that had the backing of the majority."

"Thank you, I should have mentioned the Assembly," Hecate replied. "By implication I was treating it as one of the alternatives we must consider for a future political structure. I must concede that the unfortunate turmoil at the Assembly's recent meetings has caused me to wonder whether we shouldn't suspend its proceedings for a while. And I'm not alone among the sisters in thinking so."

Rhea laughed. "Dear sister Hecate, I admire your bland assurance that things are working so smoothly that we can just continue on our present course. Regrettably, quite recent events in our midst give the lie to your panglossian scenario. Can you really ignore the fact that the chair of our board recently sanctioned the use of lethal force against two males of the Second Generation? Either or both of them might have been killed by our hand, although had it happened, our excuse would have been that their deaths were unintended. By the way, I supported that decision because we were responding to a criminal act of forcible confinement and kidnapping.

"Our little paradise is in turmoil, and you should admit the fact. I dislike the whiff of presumption associated with the Silverbacks, and I cannot agree that we should just treat what happened as some kind of adolescent prank, as you seem to be doing. By hiding yourself behind our entitlement to control the Foundation's assets, and seeking to derive our exercise of legal authority from that source, you are simply trying to avoid the key issue here, which is, to put it bluntly, one of *political legitimacy.*

"What you are doing in essence is sugarcoating the truth, namely, that the second and third generations live under a dictatorship—and it is we, sitting around this table, who are the dictators."

I couldn't resist the opportunity. "Please, on a point of

personal privilege, Madam Chair, I wish to be exempted from this characterization. As the saying goes, I was only following orders."

Hera said, "The Chair notes Marco's objection and overrules it. He is to be regarded as one of the dictators."

Gaia asked to speak. "I'm afraid I must agree with our resident devil's advocate. All together, there are, according to the latest population registers, about fifteen thousand souls residing in our territories. We can't just carry on as if nothing had happened since the little band of twelve sisters, with baby Marco in tow, decamped to the Caribbean thirty-some years ago. Soon we may turn into the equivalent of a princely ruling family in a medieval Italian city-state. We can't continue to ignore the question of the legitimacy of our rule.

"I'm facing this kind of challenge right now at Sujana University—or Solomon's House, as we prefer to call it. One of our recruits, a brilliant molecular biologist by the name of Abdullah al-Dini, is extremely unhappy with our rules against free contact with scientists elsewhere in the world. Despite the continued support for those rules by our College of Scientists, he continues to agitate and seek supporters among the staff. More to the point, he insists we have no right to make such rules."

Pandora interjected. "My position on this matter is clear, as I told Gaia. Either al-Dini stops his campaign, or he and his family should be expelled."

"It may come to that," Gaia answered, "although I'm still trying my best to negotiate a solution with the help of the president of the College. But that doesn't deal with the larger question. I think we need a firmer basis for the political authority that we now exercise."

The debate commenced, being interrupted once for a light lunch before resuming throughout the afternoon, then again for a sumptuous dinner, before continuing in the adjacent lounge with the help of various psychotropic substances far into the night. Just before exhaustion overcame them, Hera asked Hecate, as the mover of the original proposition, to summarize the emerging consensus among the sisters.

"There is general agreement on three points," she reported, having consulted her notes. "First, we must make a firm distinction between members of our tribe, on the one hand, and all other residents of our territories, on the other. There are different principles of political legitimacy applicable in either case." She looked around the room. "Agreed, so far?" Everyone nodded.

Then Artemis intervened again. "I do agree with the consensus on this point, but rather reluctantly, I must confess. Before we move on, I just wish to note for the record that our position in this regard harks back to the fateful choice we made, about twenty years ago, while we were in the Turks and Caicos Islands, to create our Second Generation. We did so for a straightforward reason: Everywhere around us, we saw humans abandoning themselves to the mad dream of bootstrapping their own existence. They were determined to use their newly discovered command over genomics in order to become a fully engineered species, not only altering every imaginable human trait in their own generation, but also passing these changes into the germline to be inherited by their children and their children's children.

"We said among ourselves that the experiment is already launched and it cannot be recalled or cancelled. We know this infallibly because we are among those who were its earliest subjects. So let's see if we can seize our fate by the throat, so to speak, and bend it to our own will. Let's see what will happen if we set free our kind to run its course on earth, as all living creatures are entitled to do. Let's see what nature's judgment will be, far into the future, when it comes to assessing our ultimate evolutionary significance among the ranks of primates."

"That's a very interesting way of phrasing what the presupposition of our actions was at the time, Artemis," Hera remarked. "What you seem to be saying is that we have a kind of commitment to the members of our tribe that cannot be shared with others."

"We took a gamble once, whether wisely or not, and we cannot revoke it now, Hera, because we performed an act of creation and continue to bear full responsibility for the result," she answered. "The rationale for it was expressed in the last of your debates with our father, as recorded in your journal. In effect what you said to him was this: Based on what modern science has told us, for the first time, about how we arose from nature and evolution, we wish to change utterly the way in which the genus *Homo* behaves with respect to the earth. We desire this because to our way of thinking what is called human history has not yet begun.

"Everything to date has been merely the prehistory of our own species, epochs ruled by animal lust, fear of the unknown, and the worship of crude idols. After all, our 'species-being,' if you will, is—supposedly—*sapiens,* those who uniquely possess knowledge. Therefore human history will begin only when, as a species, our orientation to the earth is rooted in a precise knowledge about how nature generated itself and how a self-conscious type of being emerged, spontaneously and accidentally, from earlier life forms. Simply put, the earth is our progenitor, our only mother, and we should treat her and the rest of her diverse brood with an appropriate degree of care and respect. We intend to give rise to the branch of our genus that understands itself as a product of nature and acts accordingly."

"Artemis," Hecate said, "I, for one, am pleased to hear your exposition on this point because all of the other main clauses in our political constitution follow directly from it. Now I will continue.

"Second, so far as the other residents are concerned, they are in a legal sense guests in our house—or, more precisely, guest workers. In our dealings with them we consider ourselves to be bound by the rules of natural justice, such as fairness, equity, tolerance, and compassion. Our authority to make and enforce rules governing their behavior derives solely from their voluntary acceptance of the situation. Those who cannot accept our authority are asked to leave, and in such cases we assume the responsibility to make a reasonable effort in assisting them in resettling elsewhere. But we regard our authority in this respect to be reasonable and not subject to challenge." She paused. "Agreed?"

She smiled indulgently as a different hand went up. "Yes, Ariadne?"

"Like Artemis, I wish to enter a personal note for the record. I regard the need for a firm separation in formal status between us and them as a matter of temporary strategic necessity, not as either an enduring matter of principle or a dictate of biology. As you all know, we aren't yet a fully hybridized variant, although we could become so, eventually failing to produce fertile offspring through backcrossing with outsiders, if we were to choose to remain genetically isolated from other groups of humans for a sufficiently long period of time.

"I agree fully with the account of our unique form of 'species-being' that you heard a moment ago, and I believe that our uniqueness arose in the modifications made to the neurological structure in our brains by our parents. Specifically, those manipulations produced a radical intensification of our sense of empathy by 'upregulating' the activity of certain genes. The same thing might have happened someday through a spontaneous mutation, entirely naturally, similar to what occurred with changes to the notorious FOXP2 gene, which made articulate speech possible, after we branched off from the chimpanzees.

"What we were endowed with by our parents is a very special trait, but it differs largely in degree, not in kind, with what many other humans who were not similarly engineered also possess. The bottom line is, once we are a much larger tribe with a secure foothold on the planet, we can begin openly recruiting other humans in neighboring lands, who share the goals of our kind, to join us on the basis of full political equality. I'm sure we will succeed in this endeavor. And at that point the separation between us and them will start to disappear."

"An excellent disquisition," Hera remarked, "and I fully support your concluding remarks. Unless I see other hands, I'll ask Hecate to move on to her third and perhaps most difficult summation."

"The third main point," Hecate began, "has to do with the future political constitution of our own tribe. There emerged a strong consensus to the effect that my opening proposition is seriously flawed. It appears that the devil had the better argument, which actually should occasion no surprise."

Rhea grinned and reached over to touch her. "We all know that you were asked by the Chair to set up a strawman argument, sister, and you did it well. We had to start somewhere."

"OK. The consensus is that the cozy little bunch of dictators sitting around this table—should we call ourselves for short the *bolsheviki*?—had better get its act in order, and fast. We're simply asking for trouble if we think we can just let the current troubles fester among the second generation. If we do, there will surely be another explosion, with far more serious consequences. It's our duty to think through the transition issues immediately and to present to the Assembly as soon as possible a scenario under which that body will become the ruling political organ within the tribe. Agreed?" Everyone nodded.

"All right then. I've been tasked with the assignment of laying some specific options on the table for your consideration. I've promised to do so within a month, and I'll need help from all of you to meet the deadline. If there's nothing else, I'm going to bed."

As the sisters were saying their good nights to each other, Hera asked me and Persephone to stay behind for a few minutes. "I know it's late," she said to me, "but I'd like to have a brief report from you on your progress with the site for Io's memorial at Cottonwood Spring Oasis." To Percy, she added: "You know the memorandum we've been discussing? When do you think you'll be finished with it?" Two weeks later, Hera had the following document in her hands.

THE EVOLUTION OF PLACENTAL MAMMALS
PERSEPHONE SUJANA
AUGUST 2066

Mammals are, quite simply, animals that have mammary glands. I will open this story at the point where the early mammalian line splits into two branches, namely marsupials (such as kangaroos) and placentals. The latter are also known collectively as Eutheria, a flattering term that means "true beasts." This split occurred about 125 million years ago (MYA), although it is likely that their evolutionary line, the proto-placental, stretches back another 50 million years. Placental mammals are, as their name indicates, distinguished primarily by their unique mode of reproduction, where an embryonic entity is connected inside the mother's body by means of a temporary organ—the placenta—that provides an interface across which nutrients, waste, and other materials flow between mother and fetus. These other materials include fetal cells, which remain for decades in the mother's body and can repair damaged tissue, including in her brain.

Placentals are also known for the relatively long gestation periods of their young—up to twenty-two months in the case of elephants and fifteen for sperm whales. In many species, including those as diverse as humans and badgers, the young are born in a helpless state and would perish without immediate maternal care. Among the large grazing animals we call ungulates, however, the young are able to stand and walk within the first hour of life. What cuts across all such differences is, of course, the presence of that life-giving substance, mother's milk, which is the indispensable nourishment for all young mammals.

The Cenozoic Era, a name that means "the age of new life," begins about 65 MYA and marks the appearance of the large mammals on earth. This new age originates in a series of momentous events: The extinction of the dinosaurs, probably as a result of a huge comet striking the earth, followed by a period of massive and protracted volcanic eruptions, with subsequent major alterations in the world's climate. We humans are among the direct beneficiaries of that otherwise unhappy event. For until that time, and for the preceding 60 million years, mammals had remained small insectivores, insect-eating creatures, their further evolution held in check by the unchallengeable ferocity of the ruling predators of the period, the dinosaurs.

Beginning about 45 MYA, all of the major groups of mammals we are familiar with today, such as bats, ungulates, and rodents, had come into existence. Since that time they have become the dominant form of animal life on earth. What is especially interesting is that in the intervening period, many land mammals, on continents as far apart as Australia, Asia, and North America, evolved to be considerably larger than any that now exist. All of these "megafauna," such as the immense, elephant-like Platybelodon, eventually became extinct, probably because of changing climate, starting about 100,000 years ago.

Nature is always experimenting with its creatures, of course. One innovation that replaced sheer physical size was larger brains. Compared with reptiles, mammals have evolved much larger brains in relation to body weight. But even more crucial than size is a truly radical transition to a more complex type of brain, because only mammals have a neocortex, a Latin term that means "new rind" because it's like an envelope covering the cortex. The neocortex is the part of the brain where all of its more elaborate information-processing functions occur.

The basic strategy of mammalian reproduction is to create an exceptionally strong instinctive bond between mother and offspring. This bond is forged in the first instance by the complete dependence of the newborn on the nourishment provided by mother's milk. As I shall describe in a moment, however, it is far more complex, involving emotional circuits and the work of powerful hormones in the mother's brain. The magnitude of nature's investment in the mother–infant bond is dictated by the simple fact that only the mother is guaranteed to be present when the infant is born. This overcompensation comes at the expense of the male: In only about 3% of mammalian species do we observe any important level of male parental care.

Now let's return to the fetal placenta, which I earlier described as an interface between mother and infant. It actually plays a far more active role than what is implied by that term, for the placenta asserts control over the production of hormones in the mother's body that make her ready for maternity. Cells in the placenta called trophoblasts, which are unique to mammals, regulate the production of steroids and peptide hormones in the mother's body and produce what scientists have called a "radical reorganization" of the mother's brain during pregnancy.

First, there is a dramatic increase in the level of one of the steroid hormones, progesterone, which has a sedating effect. But progesterone and its partner hormone, estrogen, also "prime" the brain's receptors for a series of other compounds, the neuropeptides—oxytocin, vasopressin, beta-endorphin, and others—making the brain much more highly sensitized to the effects of these chemicals. Finally, greatly increased quantities of the neuropeptides are produced throughout pregnancy and again at the moment of childbirth.

Oxytocin and vasopressin are found only in mammals. Both of these compounds, however, evolved from earlier peptides found in fish, reptiles, and worms, where they have important roles in sexual behavior and egg-laying. It is always useful for us to be reminded about the links that connect us far back in time with earlier and simpler life forms, for they are our predecessors, and without their evolutionary success we wouldn't be here.

Oxytocin, in particular, produces effects on the mother that are critically important to the maternal behaviors that mammalian offspring require for survival. It creates the mother–infant emotional bond, including the recognition of the infant by smell, promotes parturition, and readies the mammary glands for milk production. In herd animals the effects of oxytocin will imprint the infant's smell on the mother's brain within minutes, and thereafter she will allow only her own infant to nurse.

For the mother, the act of nursing stimulates repeated doses of oxytocin, which acts together with estrogen to flood the brain with dopamine, creating surges of pleasure. Scientists have confirmed these findings in many experiments. For example, in sheep, when ewes that are not pregnant were infused with oxytocin, they displayed full maternal behavior and bonding with infants.

We must not think that everything is sweetness and light in this story. One clue that something else is going on is the fact that, in all ages before modern medicine and in places where it is still unavailable today, pregnancy and childbirth together are the leading cause of death among human women of childbearing age. The simple truth is that the interests of mother and fetus are not at all identical—indeed, in many respects, the two are in competition with each other.

A large part of this competition is rooted in genetics. The genes of the fetus are different from those of the mother because they are a combination of maternal and paternal inheritance. The mother must balance her "investment" of resources in any particular fetus against her own needs as well as her future reproductive potential; on the other hand, the fetus in the process of development is motivated to get as much of the mother's current resources as possible. This battle is fought through the placenta because while maternal genes control embryo formation, paternal genes control the growth of the placenta, through which the mother's resources flow to the embryo.

Serious risks to the mother's health can arise as a result of this competition. Two may be mentioned for illustration. Pre-eclampsia is a sharp rise in maternal blood pressure, caused by a protein produced by the placenta and resulting from the fetus's attempt to gain more resources under conditions where the mother is facing nutritional inadequacy. Then there is gestational diabetes. Again, the placenta is producing enzymes that break down insulin, and the resulting impact on the insulin-producing cells in the mother's pancreas can cause this disease. To be fair, the adverse impacts work both ways, of course. A mother can transmit serious illnesses to the developing fetus, such as narcotic addiction, HIV/AIDS, and fetal alcohol syndrome. In short, mammalian reproduction can be a risky business for both parties.

Now let's get back to the main story. Mammalian reproductive strategies are of two types. Among rodents and similar species, large litters of pups are produced; in these cases, the mother bonds to the litter as a whole, and not to individual pups. The other strategy involves the production of a much smaller number of infants, normally from one to three, and this marks a fundamental change in the biology of maternity. The smaller numbers of offspring per pregnancy allow the development of much larger fetuses, including those with bigger brains. Whereas, in the first type, olfactory cues are the most important mechanism of maternal bonding, in the second type there is far greater reliance on completely different brain circuitry, namely the so-called "reward" mechanisms.

The second type of reproduction strategy is marked both by much longer gestation periods as well as by extended periods of infant dependency after birth. In land mammals, such as bears, and sea mammals, such as whales, these postpartum episodes, lasting for two or three years, amount to extended seminars offered by mothers to their offspring in learning how to survive. Humans, who at birth are far less fully developed than all other mammals, take this pattern to the extreme: The frontal cortex in the human brain does not complete its developmental cycle until around the age of twenty.

Many years ago one authority in this field wrote: "The evolution of mother love was essential for the evolution of intelligence." This is essentially a conclusion drawn, in the first place, from the basic description of the long gestational opportunities available to the offspring of placental mammals. Equally important, however, is what happens after birth, especially among primates. Generally, daughters remain with their mothers, that is, among their matrilineal kin, whereas males tend to leave. Scientists believe that it was in this matrilineal hothouse of intensive social interactions that the so-called "executive functions" of the human brain evolved. This is because lifelong participation in never-ending, close interactions within social groups required the evolution of strategies for coping with the stress and potential conflicts that arise there.

Negotiation and mediation of potential conflict situations among social animals is highly dependent on what is called "emotional regulation," through which inappropriate behavioral responses are suppressed and empathy arises. In support of this function, mammals have evolved complex brain circuitry, in both the neocortex and the limbic system. Studies have shown that the infant's experience with mothering, including suckling, cuddling, licking, and grooming, is essential for the achievement of a capacity for emotional regulation in the infant. Where infants were deprived of mothering in experimental situations, involving rhesus monkeys, they developed aggressive behaviors known to be related to disruptions in frontal cortex function. Young male elephants in the wild that have been deprived of mothering due to the killing of their extended female kin groups have become pathologically aggressive.

Evidence from genetic studies confirms this. As you know, we inherit one set of genes from our mother and one from the father, and embryonic developmental processes determine for each gene which inheritance prevails. Genetic research showed that genes inherited from the mother are more influential than are those from the father, precisely in the newer parts of the brain (the neocortex) and particularly in those parts involved in executive function, such as the anterior cingulate cortex.

There are behavioral types of inheritance effects as well. The amount of licking received by infant female rats determines the extent to which they will lick their own pups (these interactions are mediated by oxytocin). Among more advanced mammals, although hormonally driven responses are still important, a transition to "cognitive" determination occurs as well. There is solid experimental evidence for the importance of social learning among matrilineal kin. Young female monkeys who are deprived of normal maternal care are unable to later give such care to their own infants, and indeed are abusive to them. Among baboon females, more grooming leads to higher infant survival rates. Both male and female infant monkeys display clear indicators of depression when they are separated from their mothers for extended periods of time.

Thus the bonding between mammalian mother and infant, originally based on lactation, is regarded as the template—the necessary precondition—for all other forms of bonding in social groups, for emotional regulation, and for empathy. In saying this, however, one must be careful to avoid implying that among mammals the full burden of socialization falls exclusively on the individual mother. If we concentrate our attention just on our closest genetic relatives, we see two basic patterns.

Among chimps and gorillas, there is an extraordinary degree of mother–infant attachment, where the mother carries her infant offspring everywhere and does not ever willingly surrender possession of it for the first six months or so. But among many species of monkeys, and among Homo sapiens as well, a high degree of allomothering is the rule. Allomother refers to "anyone but mother," and the term includes relatives, siblings, grandparents, unrelated members of the social group, and fathers. Any of these can supply the constant physical closeness and attention that primate infants must have in order to develop normally; some of the females on this list can and do also provide nursing, of course. However, consistency in which other individuals actually perform the allomothering role is important to infants.

After birthing, mammalian mothers must continue to balance their infants' needs against their own: A monkey or baboon mother must continue to forage for her own food, and a human mother has her education, profession, and political responsibilities to attend to. One of the great authorities in this field, Sarah Hrdy, struck the right balance, in my opinion. She reminds us that mothers are "hormonally primed" to seek closeness to their infants, and staying close to mother is the natural first choice of the infants as well. Among most primates and indeed most mammals, this is an evolutionary truth well supported by observation and experimental evidence. On the other hand, Hrdy also stresses, for humans and for some monkeys and other mammals, that the option of allomothering is frequently relied upon under conditions where the mother can be reasonably certain that her infant will be safe and secure in others' care.

§ § §

The next time the three of us were together, Hera said to Persephone, "I loved your essay. Whenever I read anything in evolutionary biology, a strange sensation comes over me. How should I describe it? I just feel so at home in the world. We were begotten here by nature, and because we emerged out of an evolutionary sequence, almost everything about us was modeled on some earlier successful adaptation.

"I remember once reading about studies of rat behavior published in scientific journals. Although rodents and primates diverged about eighty million years ago, we're still very similar creatures. Like us, rats dream, they love to be touched and cuddled—and tickled. They can be happy or sad, they get addicted to drugs, and they know the difference between just having sex and having really good sex. They even display what scientists label 'metacognition': They're aware of the difference between what they know and what they don't know!"

Percy grinned. "Yes, I'm quite fond of rats myself, in small doses. To be perfectly honest with you, I prefer humans in small doses, too. So, why are you so interested in placental mammals just now?"

"Soon we're going to have a lot of young mothers in our midst. And at least for a little while longer, the band of sisters still makes the rules by which we live at Yucca Settlement. We sisters have formed a commitment to consider what evidence we have, furnished by scientific studies, that may be relevant to the rules we adopt here, either as laws or social conventions."

"One of the traits I most admire in you," she said to Hera, "is your rigorous self-control. You present your interest in mammalian mothers in wonderfully neutral terms. And yet I cannot help wondering whether what you are thinking about has rather less to do with evolutionary science than it has with actual recent events in our tribe."

She smiled. "Not for a moment did I think I would fool you with my crude dissembling, sister. That the sexual politics in our tribe would so quickly escalate from rhetoric and grandstanding to forcible confinement and implied threats of rape is something that took me completely by surprise, I confess. I and many of the sisters are agonizing over what, if anything, to do about it."

"Remember that only a small minority of the males was involved in the most serious episode. Most of what has been happening appears to be rather innocent posturing."

"Still," Hera answered, "although what Lateefah and Myra did was extremely foolish, we must remember that they acted in response to what they were hearing from other young men, who were not among the members of the disaffected group. They heard a good deal of attempted rationalization for the motivations of the so-called Silverbacks."

"Perhaps we were a bit naïve, sister. Did we really think that the tweaking of a few neural-circuit genes by our parents would abolish fifteen million years of evolutionary history overnight? The boys may have had delusions of grandeur in naming themselves after the dominant gorillas, but as they well know, in behavioral terms, the males of our own species are pretty much just hairless chimpanzees. Male chimps are extremely aggressive and very dangerous animals, especially when they join up into small groups. Add the technologies devised by the oh-so-clever *Homo sapiens*, and you have a recipe for constant mayhem. The single most destabilizing element in human societies, by a long shot, is young men with guns."

"Oh thank you, Percy, that's just what I needed in order to cheer me up."

"Sexual dimorphism among humans is similar to that in chimps and baboons," Percy continued. "In both cases the males easily dominate females because of both their physical size and their well-practiced belligerence. In both cases most of them won't hesitate to use violence against females. And, by the way, speaking of mothers, that includes fairly frequent cases of infanticide perpetrated by dominant males."

"So this is what we have to look forward to?"

Percy chuckled, but there was a serious tone in her reply. "Forgive me, but what was it you yourself told me just a moment ago? That we have formed a resolve to test our social rules of engagement against scientific evidence? Well, I just supplied you with some highly relevant evidence."

"Touché. I deserved that."

"Hera, we sisters have arrived at a crossroads, and our reaching it was fully foreordained on the day we took the decision to create the Second Generation. We bred a human group from the material we had on hand, which was an assortment of embryos derived from a hominin line that clearly displays its genetic inheritance from the great apes and the Old World monkeys. Males in this line are acutely sensitive to the distribution of power in social relations. They clump together in pursuit of power as naturally as blood coagulates at the site of a wound."

She made an effort to protest. "I agree that what you say describes human history to date well enough. But weren't we led to believe by our father's original scientific protocol that a heightened sense of empathy would have some kind of ameliorating effect in this respect?"

"Indeed it may, Hera, although our carefully engineered brains are fighting an uphill battle, as they say. Political power lusts for concentration, and young hominin males are the willing subjects of its lust. Evenly dispersed power is quite literally nothing, merely the latent possibility of a transfiguring metamorphosis that is at one and the same time a metastasis. Once a lowly crawler hidden harmlessly within its concealing cocoon, the butterfly suddenly soars into the sky on dazzling, brilliantly patterned wings. Similarly, power, which vanishes when distributed, but once coalesced, shows itself in the menacing beauty of disciplined columns of marching male warriors."

"A lovely metaphor, Percy. Still, I do appear to be hoist by my own petard in this little discussion. I have persuaded my sisters that our little tribe, which may or may not be a distinctive variant of *H. sapiens,* should not be in the business of experimenting repeatedly with its genome until a version is found that everyone is happy with. This means that I would be opposed to any attempt to breed the residue of male chimp aggressiveness out of the hominin line. Or alternatively, trying to breed in some of the more laid-back traits of the bonobos."

Percy was unrelenting. "Which means you have to figure out some other strategy, or else just wait around until the young men in this little tribe of ours, the one you first called *Homo carstenszi,* figure out how to reconfigure it back into the standard mold of patriarchy and male rule. They are drawn to power as moths to the flame, despite knowing that it is likely to engulf them, for its lure is irresistible."

"Some women, too," Hera interjected.

"Fair enough, some—but not all that many. If you look at the primate family as a whole, you can find clear dominance hierarchies among the females, but not a lot of violence associated with it. In most cases they seem to be able to work around the lure of power."

Before Hera could reply again, Percy grinned and said, "We should lighten up a bit. Here's a thought. Why don't we offer the heterosexuals among the boys a sweet deal? What we might call a reverse-harem arrangement? They agree to having a permanently adverse gender ratio, say one male to every five females. Their sole function will be to perform insemination functions; they'll be fed and pampered and protected from human and natural enemies. The ladies will take care of everything else. The boys won't have to bother their pretty heads with the tedious business of politics or earning a living. And no warrior stuff, except in virtual reality games."

I could tell that Hera wasn't in the mood for this kind of levity, but she tried. "Actually, we'd have to include some regular mortal combat, with opposing teams hunting each other out in the surrounding mountains. Otherwise Mother Nature wouldn't be able to keep her selection pressures on and their breeding fitness would deteriorate rapidly."

"Sorry," Percy said. "I was just having a bit of fun. Actually I think the scheme wouldn't work in any case. With that much leisure harem time on their hands, even the video games and sports wouldn't soak up all their competitive energies, and it wouldn't be long before they'd be scheming about seizing power again. I'm pretty sure that given a choice between unlimited sex and unlimited power, they couldn't resist trying to leverage the latter into having both. Even though soon after power was re-consolidated in their collective hands, most of them would find themselves back in roughly the same subjugated position as the females, kowtowing meekly to the new leaders and their thuggish enforcers. Nevertheless, I'm convinced they'd give it a try, because power is more important than sex to them."

Looking at her watch, Hera said, "I've got to go. I hope the other sisters will be more successful in coming up with solutions than we were just now."

Here I provide a short summary of the continuation of the sisters' debate that took place at a follow-up meeting of the Foundation's board about a month later.

Pandora opened with a passionate address on the history of male oppression and violence against women, concluding with a proposal that women retain forever the three-quarters majority of votes in the Assembly they presently enjoy. No one disputed the case she had made, only the proposed remedy. The group's consensus, arrived at fairly quickly, was that the basic principle of formal legal equality must be upheld, because both men and women in earlier times had put far too much effort into the struggle to achieve this.

But they also concurred it would be foolish to ignore the reality and consequences of sexual dimorphism in so many mammalian species, including our closest relatives in the genus *Homo,* chimpanzees and gorillas. The plain truth is that in all human societies to date, without exception, males have used their advantages in physical size and aggressiveness to control women's lives, and especially their reproductive capacities. The tools used have proven to be crude but highly effective ones, to wit, intimidation, physical brutality, rape, and murder. Neither smarter brains, nor religions, nor ethical systems, nor refined civilization has altered this pattern except trivially, and the improvements wrought since the onset of the democratic era remain fragile and often contested. So if the Yucca tribe must have legal equality among all its members, it must also have some other ways of safeguarding the dignity of its females.

However, some time ago the sisters had made for themselves three resolutions about further genetic engineering and reproductive manipulation in future generations of their tribe. First, they resolutely rejected the idea of seeking to alter the power imbalance between males and females by genetic engineering. They had also determined not to manipulate gender ratios, beginning with the Third Generation, but rather to allow the natural ratios to return. Third, they would allow passive screening for inherited genetic diseases, using pre-implantation diagnostics, so as to remove them from their gene pool as far as possible, but no gene enhancement.

After considering and then declining to approve Pandora's solution, they finally resolved to promulgate their own constitution in the form of a small number of fundamental laws, and to impose it on their tribe. They would create a firm barrier against the continuation of male violence and intimidation. They would deal with the fact that men's control over economic resources had always solidified the advantages they derived from superior strength. And, once and for all, they would ensure that the biological reality of the radically different roles of the sexes in reproduction would be reflected appropriately in law and custom.

Hera was given the assignment by her sisters to address the Assembly and review the crisis into which the Yucca Settlement had been plunged as a result of the actions of some of those who call themselves "Silverbacks." She began by reminding them that, beginning with the members of the new generation who were already being born, the natural ratio of gender numbers among human offspring would be maintained. The expectation is that into the future, the political structure of the Yucca tribe will be a democracy in which all citizens have equal rights and duties, and that the Assembly will remain the core of its governing structure. Then she announced a few interim governance measures as well as the three fundamental laws.

"Members of the Assembly will soon be asked to elect an executive body—a Cabinet, composed of twenty portfolios such as are found in all governments, including health, defense, economics, social welfare, and so on. These officers will be assigned to work closely with the sisters and the senior administrative personnel at the Settlement, and to participate in decision-making on all matters affecting our lives here. We propose a term of three years, after which a new group will be elected. And we have a specific request to make of you.

"As a way of addressing some of the tensions that have arisen among you, relating to gender parity, the sisters will formally ask the Assembly to elect an equal number of men and women, that is, ten each, in the first round. In return, the sisters would like to reserve the right to assign the portfolios to individuals after the group is chosen. Beginning with the second round of elections, however, the exclusive right of deciding who the successful candidates will be will revert to the Assembly.

"And now to the fundamental laws. As I just mentioned, we regard all citizens of Yucca as having an equal entitlement to the enjoyment of the fruits of our resources, just as all have an equal duty to help it prosper and to defend it against any enemies that may appear. But we cannot simply turn a blind eye to the fact that all human societies to date have imposed severe disadvantages on women and have enforced this regime by violence.

"Thus the first law is that all citizens of Yucca must undergo the same period of military training and must maintain their level of martial skills throughout their working lives. All will become proficient in weapons use and self-defense techniques and will maintain that proficiency through regular re-training programs. We wish to begin the first phase of training immediately, and for this purpose we have recruited instructors from the ranks of our technicians based in Beatty, many of whom are women and men who retired from active service in the US military. Wherever in the hierarchy of our armed forces officer ranks are thought to be necessary, a majority of the places in them will be reserved for women.

"We need to make special provisions for our women only during their years of pregnancy and early motherhood. Therefore, we will ask half of our mothers-to-be to postpone their first pregnancy until they have completed their first round of military training, which we estimate will take eighteen months or so.

"The second law deals with the matter of property. As you know, up until now, legal control over the material resources that support and protect our Yucca Settlement has been vested in the Sujana Foundation. The Foundation is legally controlled by its board of directors, composed of the eleven surviving sisters and Marco, all of whom were present when the Foundation was established in 2032 in the Turks and Caicos Islands. We believe that the directors have the moral right and duty to decide on the future disposition of its assets. However, we have been aware for some time that, as the Second Generation matures and debates matters of policy in its Assembly, we sisters must begin to yield control over the Settlement's resources. Today we announce how we will do so.

"The Foundation will issue one voting share to all of the women of the Second Generation at the time when each first becomes a mother. These shares will always carry the right of casting one vote during the periodic elections for the members of the board of directors. This share can never be transferred or sold by its bearer, but will be held in trust for all of the daughters born to that woman, all of whom together will then possess a fractional vote. They in turn will be bound by the same trust agreement, passing their collective share on to the daughters of all of them in the succeeding generation. And so on in perpetuity, down the matrilineal line, where the vote will be further subdivided in each generation according to the number of surviving female progeny.

"While the majority of the original band of sisters are still alive, all of the votes from these trust shares will elect one-half of the members of the Foundation's board. At the time in the future when there are fewer than six surviving sisters, or sooner if the sisters so decide, the holders of these shares will elect the full board.

"The third of the fundamental laws provides that no rule or policy should ever seek to limit the discretion of any woman, with regard to the fetus she is carrying, during the first trimester of pregnancy—with the important proviso that society may properly intervene, at any stage of fetal development thereafter, either to protect the interests of a fetus or to safeguard the mother's health, or both."

She then turned to the matter of recent events.

"Finally, we wish to inform everyone of the disposition of the very serious case of armed violence, kidnapping, and forcible confinement that occurred among us. Kenji has accepted the punishment handed down by the sisters' council. He will be exiled for a period of five years to our Bakersfield Hospice and the Lake Isabella Pavilion. There he will perform routine chores with the rest of the staff, but also follow a course of medical training in order to become a qualified physician. He will be permitted to make regular visits to us during our annual stay at Jalama Beach.

"Rainer, as you all know, is at present totally and perhaps permanently blind as a result of the incident at Jackrabbit Spring. Obviously his wounding was unintentional, and he has acknowledged his own responsibility for what happened; equally obviously, we regard this tragedy as more than sufficient punishment for his actions. It will take us some time to ascertain whether or not some or all of his vision may eventually be restored through one of the newer medical interventions or technologies. We will spare no expense to this end."

When I walked into Gaia's office sometime before the scheduled start time for the next meeting with Abdullah al-Dini, I found both Hera and Athena already sitting and chatting with her at the small conference table. The care-worn expressions they shared spoke eloquently of the perils they had recently confronted and surmounted, at least for the time being. But something told me that they were anticipating the onset of another and different type of battle, an intuition that soon proved to be well founded.

Bypassing the usual formalities, Hera blurted out, "We asked you to come a bit early because Gaia has something to report to you."

"Marco, you'll recall that after the last meeting in this series I was asked to arrange a chat between al-Dini and the head of our faculty association, and, of course, I did so. Because we've all been a bit preoccupied with other crises"—here the three sisters exchanged a look of mock horror—"I couldn't get the meeting arranged until two days ago. I'm afraid it didn't go at all well. In a nutshell, our new recruit reiterated his firm opposition to having any restrictions placed on his external scientific communications."

"I took the view that al-Dini should be asked to leave our territory at once," Athena interjected, "but Hera over-ruled me. Since today's meeting was already scheduled, she thought it would be fairer to wait until we had tried again to find common ground with him. But she did agree with me that you should be alerted and asked to increase the level of security and surveillance for the offices and labs to which he has access."

"I'll take care of it as soon as this meeting ends, unless you think I should leave now."

"That won't be necessary," Hera replied. "Thanks to your efforts, our various facilities are well compartmentalized and protected with separate security codes. As of now he still doesn't know anything of importance about our most sensitive operations. You can adjust the surveillance levels for his section later today." She glanced up at the monitor. "He's here."

"Abdullah, welcome back," Gaia said as she rose to greet him. "You've already met everyone else around the table, I believe."

"Yes, I have, thank you," he replied, whereupon he launched immediately into his prepared opening statement. "You'll be pleased to learn, I hope, that I've been following up with some reading in the sources you referred to at our last get together. I particularly concentrated on the 1945 Franck Report and its discussion of the responsibility of scientists. I wrote down a few of the key passages so that I could refer to their exact wording today." Here he paused to extract a sheet from his briefcase.

"This is the key sentence, where Franck and his coauthors write: 'In the past, scientists could disclaim direct responsibility for the use to which mankind had put their disinterested discoveries.' Then they go on to argue that the development of the atom bomb changes the situation in a fundamental way, requiring what they refer to as a more 'active stand' on the part of scientists. In other words, the discovery of nuclear energy places a dividing line across the history of science. They appear to be suggesting that these events mark a permanent alteration in the way in which scientists should relate to the rest of mankind."

Gaia smiled broadly. "I agree with your choice of text, my friend. It is indeed a landmark document, written at a perilous time in modern history. And I'm delighted that you focused on what is undoubtedly the key sentence therein."

"To be sure, in what actual ways scientists should act differently, as a result of their changed situation, isn't entirely made clear there," al-Dini added.

"True enough, but what we are left with are two very important ideas, both of which can be inferred from the main statement. The first is that after the nuclear bomb, scientists must accept 'direct' responsibility. I interpret this as meaning *personal* responsibility. Science can appear to be a disembodied process because it normally constructs its decisive breakthroughs using many hands, in many places, step by step, often with major missteps and wrong turns along the way. Yet this vast enterprise is, in the end, nothing other than the collective result of all the individuals who labored in its service. And when each phase of their work is finished, what they hand over to the rest of humankind is some form of enlarged technological powers."

"I'm not certain I agree entirely with your interpretation on this point, Gaia, although that may be of little consequence in the end, because it's the second half of their sentence that delivers the punch. There they claim that scientists bear direct responsibility for the uses to which *others* put their discoveries. I must confess that I was deeply troubled by this idea, and after much reflection I have concluded that I am unwilling to subscribe to it. Nor do I think that very many of my fellow scientists would disagree with my position."

"May I inquire as to your reasons?"

"If I were to condense them into a single proposition, it would be this: No one can anticipate all of the uses to which any breakthrough in the scientific understanding of nature eventually may be put. And even if some objectionable uses were to be anticipated, which isn't hard to do for any of us, no scientist has ever been, or ever could be, in a position to guarantee that they would never be realized by someone, somewhere, sometime. There can be only one result from adopting this doctrine, and that is, to renounce the practice of science."

Before Gaia could reply, al-Dini spoke again. "And, may I add, to renounce not only the practice of science, but of all intellectual life. The fruit of any philosophical or literary or artistic work, say, that represents for its author the search for a different kind of truth, may be perverted later by others who use it to advance an evil cause. In any event, no artist or thinker can provide to his audience an ironclad guarantee that such a thing will never happen, even far into the future. So his only recourse would be to exorcise all ideas from his mind and to abstain completely from the process of intellectual creation."

Athena leapt into the fray. "No, you cannot legitimately broaden the scope of the charge so greatly in the hope of discrediting it! How can you draw an analogy between a colossal weapon, such as the hydrogen bomb, and a novel, or philosophical work, or a painting, sculpture, or piece of music? That's absurd! None of those other products of mind could ever carry the potentiality of such destruction. The fact that some Nazis claimed to draw inspiration from the works of Wagner and Nietzsche added exactly nothing to the horrors they perpetrated against their victims. This line of argument simply obscures the key issue, which is the escalation and accumulation of power. Modern science and the technologies it begets combine to forge the most fearful ring of power it is possible to imagine!"

"Ring of power?" al-Dini asked. "What do you mean?"

"My sister is in the habit of speaking in metaphors, Abdullah. Please forgive her. Athena, would you mind rephrasing your point?"

"Sorry. But my proposition is quite a simple one. There is something fundamentally different about the intellectual activity that we refer to as 'modern science.' It differs as much from most of what passed for 'science' in preceding centuries as it does from literature, music, and all the other arts. What differentiates it is precisely its single-minded orientation to uncovering and reproducing 'the way nature operates.' This isn't my idea; the English philosopher Francis Bacon explained it all some four hundred years ago. Figure out gravity, and eventually you'll be able to set out on interplanetary voyages. Array the periodic table of elements correctly, and you can manufacture all kinds of useful things that nature never got around to making. Find how genes work, and you can change the traits of plants and animals to your heart's content."

"It's not quite as easy as you make it sound, Athena."

"Of course it isn't easy, Abdullah, and by the way I didn't claim it was. Nor do I mean to slight one bit the scope and brilliance of the intellectual achievement that is involved. When, following Bacon, I say that at its core modern science expresses in a disciplined and methodical form the human 'will to power,' I do not intend thereby to deliver a backhanded compliment. This is simply what all of modern science wishes fervently to do. Inherent in its method of inquiry is a 'readiness' to translate knowledge into operational power. The ability to manipulate is meant to follow directly from every act of understanding, whether we're talking about organic chemistry, molecular biology, or nanoscience."

"The purpose of manipulation is, quite simply, to

devise techniques and products that will improve human life. There is no other agenda for a scientist."

"And so we're right back where we started in this conversation. When confronted with techniques and products of quite another kind, the ones that make human life more miserable and precarious, you just wash your hands of the whole matter and say, 'Very sorry, but that has nothing to do with us. Please address your complaints to your governments.' If I may say so, it seems just a little self-serving to take credit only for the good stuff."

Hera stepped in just as he prepared to deliver his riposte. "You may be surprised to hear me say this, Dr. al-Dini, but I for one am prepared to concede the point you made earlier. If we don't frame the issue before us with great care, we risk getting mired in the type of absurdities you have so eloquently elucidated for us. I would like to try a different tack and see if I can rescue from scorn the apparently scandalous notion you and Gaia were debating. First, to restate it: 'Scientists must bear direct responsibility for the uses to which mankind may put their discoveries.' Is that a satisfactory shorthand version?"

"Indeed it is. However, before we continue I must be permitted to address Athena's charge. What is actually done in the end with the discoveries of scientists is outside their control. It is not within the means, or indeed the abilities, of scientists to superintend what happens with the fruits of their research. In all technologies, good and evil purposes are inseparable; only governments have the right and the capacity to disentangle them. Indeed, they have adequate power and authority to shut down the scientific enterprise itself entirely if they so wish. But to expect scientists themselves to do so is simply absurd."

"I believe you've made your position crystal-clear," Hera answered. "My sister Athena mentioned Francis

Bacon. Are you by any chance familiar with any of his writings? In particular, do you know his utopian fantasy, *New Atlantis*, which he wrote around the year 1625?"

"I've heard of him, of course, and I'm aware that he's regarded as an important early champion of a new kind of science. If I'm not mistaken, he also called for government support and funding for scientific research, a remarkable act for his time. But I've never heard of the book you mentioned."

"I, too, have brought along today a short passage, this one from the work I just mentioned. Let me briefly set the scene before I read it to you. More than four centuries ago this writer imagined an institute made up of scientists and innovators who lived and worked entirely independently of any external control. However, they were strongly motivated to make discoveries that could be turned into new technologies, bringing concrete benefit to people's lives. You may recall that at your first meeting, Gaia mentioned a funny name, 'Solomon's House, or the College of the Six Days' Works.' That's the name he uses for his scientific society in *New Atlantis*.

"I still don't understand the name, but in substance his vision sounds very much like what a modern university actually is," al-Dini said. "A truly prophetic idea."

"Indeed it is, quite remarkably so. But Bacon was not only a visionary; he had a highly practical mind as well, and he had long experience of high office in government. He was also intimately familiar with the alliance of church and state, which in his opinion often colluded in throwing obstacles in the way of progress. So when he penned his little utopia, he was careful to bestow upon his imaginary university and its work complete freedom from external control of any kind, secular or religious. Here is his description of that institution, in his own words:

> And this we do also: we have consultations, which of the inventions and experiences which we have discovered shall be published, and which not: and take all an oath of secrecy for the concealing of those which we think fit to keep secret: though some of those we do reveal sometime to the state, and some not.

"You will see at once, I'm quite sure, the relevance of that passage to the conversation we are engaged in today, in this room."

After pausing, al-Dini asked for the sheet on which the quotation was transcribed and studied it for several long minutes. Then he chuckled and said, "You have chosen well. You say that was written around 1625?"

Hera nodded. "You see, my sister Gaia has been careful, in the course of these little discussions with you, to keep one most important point clearly in focus, a point that this passage from Bacon makes very eloquently. Namely, this: A reorganization of scientific activity along these lines can only be carried out with the full and free cooperation of leading senior scientists. How else could it be conceived? Oh, I suppose one could invoke the threat of the concentration camp, on the Hitlerian or Stalinist models; but besides offending every human value we hold dear, I just don't think such measures would be very productive, certainly not in a way that could be sustained over a long period of time, which is what scientific progress requires."

"And yet your model most certainly does violate at least some important values, Hera. Notably, unrestricted freedom to publish, full and free exchanges with fellow scientists around the world, and freedom of movement. What you aim to construct is a beautifully gilded prison—I've observed its attractions first-hand ever since I arrived here—but it is a prison nonetheless. And sooner or later it won't even fulfill its purpose. Eventually the engine of knowledge generation will seize up because it has too narrow a base of operations. A defense of orthodoxy and protection of reputations will arise and strangle creativity, and scientific progress will grind to a halt."

"Again, you may be surprised to learn that your alternative scenario concerns me greatly because I admit that things could turn out as you describe. Every choice we make has its own set of risks, its own upside and downside. Ironically, what I'm counting on, in order to hold at bay the real dangers you've identified so clearly, is the 'ethos' of science itself. After four hundred years of continuous development, both the spirit and practice of scientific inquiry is deeply imprinted in the minds of its practitioners. Every one of them is strongly motivated to uphold the integrity of research and the questioning of received wisdom. There are other safeguards, too. Our operations are already so large that most of our academic fields have been divided into two or more independent departments, housed in separate locations; they compete against each other for grants and resources."

"Be that as it may, I still think you're making a big mistake. And despite the time we've spent in these discussions, I still don't even understand why you think what you're doing is necessary."

"Perhaps the reason is that we haven't quite finished with our exposition," Gaia broke in. "However, I will promise you that we can complete the series today, so you don't have to fear becoming trapped forever in a nightmare from which you can't seem to wake up. Can I persuade you to indulge me on those terms?"

To my eye he didn't look thrilled at the prospect, but he gave his consent.

"We've already picked up the thread of discussion, thanks to you, when you recalled the words of some physicists at the time the atomic bomb was successfully demonstrated. For them, as you reminded us, with the discovery of nuclear fission, scientists had crossed into an entirely new and fateful zone. And, of course, we know now that, once having entered, we find there is no going back, and no exit at the other end either. Although the world has avoided the worst catastrophes, including a global 'nuclear winter,' we have seen nations fire nuclear-tipped missiles at each other, with horrendous casualties, and we live even today with the constant threat of having some dissident group detonate a 'dirty bomb' in support of its cause."

"Perhaps we should just round up and shoot all the physicists, and thereafter all would be well," al-Dini muttered *sotto voce.*

Gaia ignored the remark but looked directly at him. "Then it was molecular biology's turn, and the whole relation between science and 'the people,' if I can put it that way, changed dramatically."

At the mention of his own field, al-Dini was suddenly more alert, as evidenced by his accurate anticipation of her line of argument. "Almost certainly you mean that science's discoveries and applications began to hold direct and immediate interest for the average individual, especially in terms of medical therapies."

"Precisely. But I would go further and even dare to use the word 'intimate.' I don't think I exaggerate in the slightest in saying that for all human cultures, the same few domains occupy a place of both great sensitivity and fierce—indeed, violent—attention. In a nutshell, these are sex, gender, reproduction, and family, and, lurking deeper in the collective mind's recesses, the systems of morality that steer behavior with respect to them. It's no accident that all religions seek to appropriate these domains and position their priests to be the sole authoritative interpreters of what is right and wrong with respect to such matters."

"I'm not sure what you're getting at, Gaia. I would have thought you'd emphasize the direct, personal benefits to people's health resulting from the better understanding of biological processes. Look at the advances we've made in eliminating so many truly terrible inherited diseases through genetic screening. Look at the enhancements to mental function we've been able to achieve, in terms of memory and cognitive function in old age, as well as the better treatment of mental disease. Look at the scope of the choices many people now have when they're planning to have a child."

"I think I know what Gaia is getting at," Hera noted. "It's one thing to understand what makes the stars shine and the earth warm, or to change the characteristics of plants and animals in order to make them more useful to us. But it's quite another thing altogether to present the human being itself as simply an evolved biological system, an organism that is, in every respect, just like every other plant and animal, and ask: 'Now what parts of this system would you like us to change for you?' And: 'Would you like us to make some of those changes in the germline, so that your own offspring's children will inherit them, too?' I think you may agree that those types of manipulations are just a little bit more sensitive than, say, figuring out how to sequester carbon dioxide underground in order to combat global warming."

"Of course; they are highly sensitive interventions! That's exactly why people are so delighted with what we can offer them—because such things as having healthier minds and bodies in their children matter more than almost anything else. What you said is true; the contributions to people's welfare that have been made by chemistry and physics are largely invisible. People see the technologies and the products in everyday life, but most of them haven't the faintest idea about their connection to the scientific discoveries that made them possible. With biology and genetics, it's different, I'm pretty sure; people are acutely aware that the therapeutic interventions we develop are a direct result of our new understanding of how their own bodies work."

"And their minds as well? Is there anything more sensitive and intimate for each of us than our own mind?"

"Indeed the same proposition holds there. As you know, the focus of my own work is on gene function in the brain. The more technologically advanced societies have long been substituting 'brainpower' for physical labor, and they also have highly competitive economies. As soon as word spread that neuroscientists were hunting for ways to enhance traits such as memory, motivation, attention, and cognitive performance generally, many were lining up to experiment on themselves with the therapies, often before either safety or efficacy had been established.

"It didn't matter to them that these therapies were designed to overcome deficits in brain activity related to disease, accidental damage, aging, or gene malfunction. Long ago, neuroscientists came to the conclusion that they had neither the right nor the power to deny the demand made by healthy people for cognitive enhancement. It was even given a name—cosmetic neurology, by analogy with cosmetic surgery. Today it's a huge and very lucrative business in some places."

Hera smiled at him indulgently. "In a way you've made my point for me just now. When people perceive the personal benefits of such interventions, and have the resources to create effective demand for them, it's clear they're not going to wait. Life is a moving escalator that cannot be switched off. What's called 'discounting the future' tells us that benefits realized in the future are worth far less than immediate gratification. Demand intensifies as wealth and disposable income increases, so the pace of innovation accelerated dramatically, beginning in the last quarter of the twentieth century."

"For the great majority of scientists those developments were most welcome. They had a far better chance than their predecessors to see their discoveries turned into useful products and technologies within their own lifetime—and sometimes to make some serious money for themselves in the bargain."

"I believe a longer perspective can be helpful at this juncture. Are you familiar with 'Ötzi the Iceman'?"

"Of course. He's the ancient human figure whose remains had been preserved in a glacier in southern Austria. If I remember correctly, he was discovered toward the end of the last century when the glaciers were melting, and through radiocarbon dating it was established that he flourished something like 5,300 years ago. Why do you mention him?"

"Just to remind ourselves of the phenomenal speed of human technological advance, I suppose. Ötzi had a bow and bone-tipped arrows, animal-hide clothing, finely constructed footwear, a cloak of woven grass, a beautiful copper axe, a knife, pouch, materials for starting fires, and other small items. There is evidence of the use of plants for medicinal purposes, and in addition to game, his diet included einkorn, an early form of cultivated wheat. In effect, that's a summary of the technological state-of-the-art for his time. But five thousand years is the blink of an eye in evolutionary terms. The pace of change since then is nothing short of astonishing."

"As I said, we all benefit today from the speed at which innovation proceeds," al-Dini remarked. "Ötzi and his fellows may have been resourceful people, but few among us would want to live like them. And despite the fact that from our perspective their technologies were rather primitive, he still managed to get himself murdered with one of them, as I recall. So little has changed in that regard at least."

"I believe my sisters and I would beg to differ with you there, Dr. al-Dini. Like those atomic physicists we referred to earlier, we think that humankind did 'cross the Rubicon,' so to speak, when its technological capabilities arrived at the mastery of nuclear fission. In a very different sense, what molecular biologists accomplished not long thereafter was decisive because it meant that there would be no turning back and no slowing down. As we just discussed, this happened because biology and genetics 'personalized' science for the people. Now they could imagine the possibility of ordering up tailor-made accessories for their bodies and minds. And, of course, others with different motivations could imagine a host of exceedingly wicked new horrors that they might inflict upon their enemies."

"I knew that sooner or later you'd get to the downside, Hera, and insert into this conversation the problem of escalating risks. Who can deny the horrors that have already been unleashed, including the use of genetically engineered pathogens and dirty bombs? Who can deny the evidence telling us that here and there around the globe, fanatics are at this very moment trying to tweak the genomes of virulent viruses and bacteria so that we have no defenses against them? One of them is reported to have said on a jihadist website that his biological weapon will make the Black Death look like a garden party by comparison."

"Well put. The doctrine of martyrdom is especially pernicious because the perpetrators care not a whit if they themselves should wind up being consumed in the conflagration they designed for others. The idea that one could be rewarded in paradise for all eternity for causing, say, hundreds of millions of their fellow humans to suffer an agonizing death is, to my way of thinking, one of the most remarkably perverse manifestations of the religious mind."

"What does this all mean except that the rest of us must fight them with every weapon at our disposal? I had a pretty good idea from our last discussion where your ideas were leading. After studying the Franck Report, I had a vague recollection of having read something else long ago relevant to its theme. And I found it easily, thanks to the miracle of digitized information storage and search."

He pulled another single sheet of paper from his briefcase. "Here is the exact wording: 'There are two ways of dealing with dangerous technologies. One is to keep the technology secret. The other is to do it faster and better than everyone else. My view is that we have absolutely no choice but to do the latter.' Those few sentences sum up perfectly my own attitude on the matter. I hope to persuade you to adopt this perspective yourselves."

Hera glanced at her sisters and chuckled. "I think you may have underestimated us by a wide margin, Dr. al-Dini. I'm not in a position to give you any details, but I will inform you that for some decades we've been collaborating on a joint venture with the US government. One of our objectives is to try to anticipate the development by others of engineered biological agents that are designed to be weaponized. We have response strategies in place for types of pathogens that no one else has even dreamed of yet."

He actually leapt from his seat. "There, you see? You do agree with me after all! This is excellent!"

"Well, actually, yes and no. Our main objective is to protect our own personnel, and we think we have a good chance at achieving it because we occupy a fairly small space that is, as you know, entirely secured against unauthorized entry. We have good systems of sensors and defenses in place against airborne agents. But given the state of chaos on this continent, I have no idea how successful the US government will be in protecting its citizens. As for the rest of the world, take a guess."

"Still, what's important is the philosophy that lies behind what you are attempting to do. You want to be out in front of everyone else and always to confront your enemies from a position of strength. This is the thrust of the remarks I just quoted to you. Obviously, based on what you yourself just told me, you agree with them."

"On the contrary, if I may speak frankly, I think it's quite a stupid idea."

He slumped backward in his chair, crestfallen. "I don't understand."

"What we do within our own operations is an attempt to protect ourselves from a particular type of threat, and I pray that we will succeed. But, in general, the proposition you advance is, in our estimation, reckless in the extreme and excessively risky. The problem we have with it is that you, and those who think like you, have neglected to evaluate fairly the other option. You dismiss it without even giving it a hearing."

"Your alternative is to keep what you know hidden away. But it just won't work. There's plenty of evidence to support my view, notably in terms of atomic secrets."

"That's a practical objection, not a statement of principle. Are you saying, in effect, that if we *could* figure out how to impose a blanket of secrecy on further advances in science, and thus on all the potential applications of this new knowledge, you would support the alternative?"

I'm not an unbiased observer here, but it did seem to me that al-Dini was momentarily stumped by the question. Finally he said, "No. You have been frank with me, so may

I return the favor? Why should I trust you? I think there's a good chance that under your 'blanket of secrecy' you will, whether intentionally or not, end up by shutting down scientific research. For myself, I can only have confidence in its future if the enterprise remains out in the open, where we can all be assured by the evidence before our eyes that the work goes on."

"I agree you have no reason to trust me. But let's try to move away from personal judgments, shall we? Phrased in more impersonal terms, we are simply evaluating differently the risks and risk-benefit trade-offs of two different solutions to a problem. Where we agree, I believe, is in acknowledging the seriousness of the risk scenario. I think you might even concede that the risks posed by dangerous technologies have risen exponentially during the past one hundred and fifty years or so—exactly in proportion to the scope of the powers for manipulating matter and energy that science has bestowed on us. You appear to believe that institutions, such as governments, international agencies, treaties, and so forth, continue to allow us to manage those risks reasonably well. Do you hold such a belief, in fact?"

"Not exactly. I suppose I hope that such is the case, although I claim no expertise in the study of the strengths and weaknesses of institutions. My claim is a more straightforward one: No matter how great the risks in question, shutting down the free exchange of scientific ideas cannot possibly be regarded as a sensible response to the situation. New scientific investigations offer us the only hope of keeping one step ahead of the malcontents and terrorists. I have already explained why I believe your solution is likely to bring such investigations to a halt."

"I'm not ready quite yet to give up trying to persuade you otherwise. Let me offer two examples of real events that are, I'm sure you'll agree, directly relevant to our theme. The first dates from the 1940s, and again we are back with the atomic physicists. As work progressed in New Mexico on the bomb, Edward Teller speculated one day that setting off such a weapon might initiate an explosive, uncontrolled chain reaction throughout all of the nitrogen atoms in the atmosphere of the entire earth. Mention of this possibility reached bureaucrats in Washington, D.C., who, of course, freaked out. Hans Bethe, who headed the Theoretical Physics Division at Los Alamos, was given the assignment to analyze this risk, and a group working under him, including Teller, calculated that this could not happen; the fear was allayed, and the project proceeded to a successful conclusion.

"But just try to imagine the drama! World war is raging, a large team of physicists living in a remote New Mexico town—many of them refugees from European fascism—is rushing ahead frantically with experiments and calculations in an entirely uncharted area of their discipline, all the while fearing that Nazi Germany will beat them to the punch. The entire project is, of course, shrouded in great secrecy. Then suddenly, the specter of a catastrophic global risk is raised. Some senior scientists calmly examine the issue and, on the basis of some twenty pages of equations and calculations, conclude that the risk is actually nonexistent."

"They were correct, of course," al-Dini added.

"Yes, but the question before us is how they could justify taking any such risk at all. The context in which they were working gives us the answer, as we already agreed during our prior conversation. The countervailing risk was staring them in the face—Germany might succeed. They could imagine Hitler in his bunker, fantasizing about 'wonder weapons' that would bring ruin on his enemies as he went down to defeat, and they would have known that a theoretical risk of global catastrophe might have added to the atomic bomb's appeal for him.

"So the Los Alamos contingent assumed the responsibility to decide, on behalf of humanity as a whole, that they could trust their own judgment about the possibility of atmospheric disintegration. And personally, I think they were right to do so. Under these specific circumstances they were justified in pushing ahead with the development of this fearsome new technology."

"Good. I agree. You have another case to relate?"

"I do. This one is a little episode that occurred toward the end of the last century. A panel of distinguished American physicists was convened in 1999 at the request of the science advisor to the President of the United States. They were asked to review a proposed experiment to be carried out at the federal government's Brookhaven National Laboratory, using a device called a relativistic heavy ion collider. The reason for the review was a claim by a participating physicist that the experiment might create a black hole in the vicinity of the earth that could swallow the planet whole, instantaneously. The panel concluded this was very unlikely to happen, on both theoretical and empirical grounds, so the experiment was allowed to proceed."

"This one is also in the realm of physics, which is far from my own expertise. However, it sounds similar to the first in that the purported danger itself turned out, on closer examination, to be purely hypothetical. Thus this risk, too, was shown to be nonexistent."

"Not quite, actually. I looked into some of the literature on it published later. Remember that we're talking about the risk of annihilation for every one of the world's present inhabitants, as well as the accumulated products of all civilizations to date. We must consider also what the economists would refer to as the 'opportunity cost' of sacrificing all of their future descendants, and their achievements—including their scientific discoveries!—for all time to come. Finally, of course, the external compulsion represented by an implacable foreign enemy was absent here. When you put the issue in its full and proper perspective, the calculation of risk acceptability in this particular case becomes quite fascinating.

"For the sake of argument, please just accept my word that the risk was not considered to be nonexistent. Some of the authorities who reviewed the issue characterized the risk as being very unlikely or 'negligible'—in technical terms, perhaps one chance in a hundred million, which is about the same annual risk we all face of seeing a large asteroid plow into the earth. Now, I put it to you: How much expected benefit would you require the experiment to yield in order to accept a risk with such a combination of likelihood and consequences?"

"I hesitate to answer without having the opportunity to examine the documents in this case. It's still possible that further analysis would show the risk to be not vanishingly small, but really nonexistent—in other words, based on a misunderstanding of physical laws."

"Perhaps. But could you and I agree on this: If we did not know, with a very, very high degree of confidence, that the risk in this case was indubitably zero, that is, really and truly nonexistent, there would be no level of expected benefit that could justify our allowing such an experiment to be mounted."

Fearing a trap, he allowed himself a good minute or so of silent contemplation before replying. "All right. I don't want to be seen as being simply obstinate. I will agree."

"Good. By the way, I'm not playing cat-and-mouse with you, Dr. al-Dini. These issues are far too important for us to be engaging in mind games. You have just agreed that some scientific experiments that may be proposed may also be forbidden, not arbitrarily, to be sure, but as a result of a rational risk-benefit calculus. It follows that someone, or some group, must have the power and authority to make such a decision. Those who claim such an authority must not be in a conflict of interest, of course. Specifically, those proposing the dubious experiment, and any other directly interested parties, must be excluded from the review of it."

"I do concede that there must be an appropriate authority involved, and that conflict of interest must be avoided. However, my view would be that only the community of scientists, in the broadest sense, which would include experts in risk-benefit analysis, should be entitled to make such a decision. It should not be entrusted to bureaucrats and certainly should not be taken on the basis of a public opinion poll."

"Well, I may want to quibble just a bit with your last comment. The ordinary citizen might reasonably believe that she was entitled to participate in the analysis of risk, in a case where the downside entails the instantaneous vaporization of the earth. But let's leave aside that little technicality for now. You mentioned that the community of scientists is the only legitimate arbiter for such proposed experiments. Here, too, I might niggle, and ask you to identify what mechanisms scientists have put in place, on a global basis, to implement your rule. Let's leave that aside as well. What you have proposed sounds to me as if it would satisfy Francis Bacon's criteria for how his community of scientists, Solomon's House, would operate."

At the mention of Bacon's name again, al-Dini allowed himself a broad smile and a hearty laugh. "You swore to me that you weren't setting any traps, but I seem to have stumbled into one anyway. I compliment you. I also surmise that you aren't recalling a four-hundred-year-old idea merely to display your erudition. You mean to implement this idea here and now, don't you?"

Hera, too, smiled. "Yes, exactly so, that is our intention."

"By whose authority do you presume to act, if I may ask?"

"We lay claim to a kind of moral authority, Dr. al-Dini. Our Sujana Foundation sponsors the research university here, providing the lion's share of its funding. With few exceptions, such as the microbiological labs specializing in pathogens, and the innovations used in our medical supply firm, the Foundation has no pecuniary interest in the outcomes. Most of the work done here is pure and applied research that is, as they say, curiosity driven. Almost all of our scientists pursue self-directed inquiries. We sponsor it because we value the pursuit of science for its own sake, just as you do."

"Your remarks don't really address my question. Surely moral authority does not inhere in a group just because it pays the bills."

"As you well know, without money there's no research. But you're right, that's not the basis of our claim. It derives instead from our commitment to protect the autonomy of scientists and the integrity of the scientific method. We propose to break the chain further along, at the point of public dissemination and open-ended innovation. We ask our scientists to accept the need for these measures and to cooperate with us in enforcing them. They do so voluntarily."

"I've been discussing these matters with your group here for many hours already, and I'm still not clear on your justifications. Perhaps that's because we keep getting diverted into little back-channels with stories about your scientific heroes."

She laughed. "There may be one or two more such stories before we're finished here, providing you will indulge us. As for our justifications, they aren't particularly complicated. We don't think that humanity in its present state is fit to manage the risks of dangerous technologies.

They just can't be trusted with the extraordinary set of new powers that modern science has bestowed on them. The religious mind, in particular, too easily assumes pathological forms, seizing any means at its disposal to wreak havoc in the world. Above all, we fear that science itself will fall victim to the fanatics because they hate, with a murderous passion, the evolutionary story of creation. We're not prepared to take the risk that sooner or later they will succeed. So we intend to hide it away and keep it alive in secret."

"And you've decided all by yourselves that you have the necessary qualifications to carry out this bizarre scheme?"

"I suppose we have. But our inspiration comes directly from some of the most revered figures in the history of science, particularly Max Born and Albert Einstein, who were tormented by what had happened in the uses to which science had been put in their lifetimes. Also from the heroic figure of Leó Szilárd, who has been mentioned a number of times in these conversations. He conceived the idea of a nuclear chain reaction in 1933 at the time when he had arrived in England after fleeing Germany. And as soon as he did so, he began to fear what would happen if Nazi Germany proved capable of developing this technology into a weapon.

"Over the next ten years he watched nervously the further evolution of atomic physics and, as I mentioned previously, sought repeatedly to persuade other leading scientists, such as the Joliot-Curie team in France, not to publish the results of certain crucial experiments. He achieved limited success, by the way. But his acceptance of the notion that scientists must accept responsibility for what happens with the potent knowledge they generate inspires us to try to finish what he started."

"So you intend to hide it away. For how long?"

"The five thousand years or so that separates us from Ötzi is, as I said, a mere blink of the eye in terms of evolutionary timescales. So my answer to your question is, it all depends on how long it takes for humanity to be ready for the types of powers that modern science unveils. Maybe a hundred years, but maybe a thousand, five thousand, ten thousand, or even longer. What does it matter? If we play our cards right, and avoid the great waves of extinctions that have afflicted plant and animal species in the past, we've still got a billion years to go, in round numbers, before the earth's death cycle begins."

"That strikes me as being a bit indifferent to the welfare of others. Just consider all the suffering that could be avoided by allowing the steady applications of scientific discoveries over such a period."

"Once again you focus purely on the upside. This is a risk calculus, Dr. al-Dini. Your strategy is to out-compete the bad guys, but don't you see, you're creating a scenario in which everyone on the planet faces a state of continuously escalating risk. We, on the other hand, sense the danger that all of advanced civilization, including the legacy of science, will one day be sucked into the maelstrom and disintegrate. We want to bring the escalation to a halt. With the help of others, we'd also like to try to withdraw a lot of what's already out there in the public domain by consolidating scientific research inside a few institutes like ours."

Gaia added, "To rephrase something Athena said at our first meeting, we intend to protect the future of science by severing its connection to industry and the state."

Athena intervened before al-Dini had his chance to do so. "By the way, I don't accept the charge that we're indifferent to the suffering that could be alleviated by the prompt application of additional scientific discoveries.

More science, right away, just isn't the kind of succor that the modern world needs most urgently. Far more effective would be, for example, a decent minimum standard of living for the poor, drastic cuts in consumption levels in advanced countries, control of nation-states through international law, less environmental degradation, and a gradually decreasing population level everywhere in the world. *None* of what's most urgently needed requires the development of new technologies. I repeat: *None.* In fact, with a few exceptions, nineteenth-century technologies would be more than adequate for the task."

"Then let me withdraw my comment and apologize to you, Athena. I intend no insults. I am not as unsympathetic to what you're trying to accomplish as it may seem from my remarks so far. I've tried to test your propositions with my questions, and I must say in all frankness, you mount a spirited defense. My new colleagues at the university here have assured me repeatedly that your Foundation is committed to the protection of scientific integrity. I'm willing to wait and see whether everything works out to our mutual satisfaction. And now I should go and attend to business in my lab."

He rose quickly, shook hands all around and headed for the door. The rest of us were left with a variety of bemused expressions on our faces as we contemplated this abrupt termination of our little colloquy.

"What was that all about?" asked Athena.

"My impression is that he's had quite enough of this friendly banter," I volunteered. "He can see by now that he's not going to dissuade you from following the course you've charted. My guess is that, despite what he said upon taking his leave, he doesn't like the situation, but he hasn't yet figured out if there's anything he can do about it. And there may be another factor at work as well."

"Which is?" Gaia asked.

"The scuttlebutt among the university staff who have gotten to know him is that Dr. al-Dini is experiencing some anguish over reporting to females. And apparently the fact that none of you wears the hijab doesn't help."

As we disbanded Hera fixed me with a baleful stare and said, "Watch him."

Part Four

The Childhood of Humankind

The flowers grow pale in the twilight.
The earth breathes deeply of quiet and sleep.
All yearning now wills itself to dream,
as weary people go homewards,
to find forgotten joy in sleep
and to learn youthfulness anew.
Mahler, *The Song of the Earth,* "Farewell"

When I took Ciso to meet the information technology sisters, Ariadne, Themis, and Moira, ensconced in their cool underground cave inside Yucca Mountain, I was astonished at how large their staffing complement had grown, and I said so.

"That's because you've been neglecting us, Marco," said Ari merrily. "You haven't been around since we became an official technical support site for the Internet, which you probably do know is now under a United Nations mandate."

"No, it's because I have nothing to offer you in the way of expertise. Except that I did help to plan the local security for the gigantic server farm you just had installed."

"Right. The UN officials consider our rather remote and well-protected setting in the dry desert, which of itself treats all machinery with such loving care, to be ideal for their purposes. We are designated as one of the primary switching nodes for the Americas. Having our own proprietary satellite in a geosynchronous orbit doesn't hurt either."

"So what do you do for excitement?"

"Well, I must confess that identifying and taking down jihadist and other terrorist websites can be pretty stimulating at times. All of the cooperating nodes in the UN network gang up on them as soon as they've been spotted. We have automated search algorithms, of course, which makes finding them a lot easier than it otherwise would be."

"Are you looking for anything in particular?"

"Mostly recipes for building advanced weaponry, especially those using bioweapons and radioactive substances. And any basic scientific information that might aid such objectives. So sometimes we have to persuade legitimate scientists and research institutes to take stuff we're worried about off their websites."

"What if they tell you to bugger off?"

"A short, sharp denial-of-service attack seems to get their attention promptly. Then we re-contact the site sponsors to see whether they've become more inclined to cooperate with us."

"I'm sure that most are appreciative of your efforts. But listen, I didn't come here just to shoot the breeze. My valued associate Ciso and I need your help. We want to outfit him with the very latest in miniaturized electronic tracking gear."

She sized up my companion. "What's the matter with him? Does he have a tendency to get lost?"

"Yes, sometimes, especially when inebriated, but that's not the reason. He's been solicited with an offer of substantial rewards in return for supplying the other party with some bits of information. Nothing very sensitive, you understand—merely our site layouts, surveillance gear, passcodes; simple stuff like that. We've decided to play along, at least until we can figure out who's behind the inquiries. So we want to keep very close watch on our friend here during his encounters."

Ciso had first been approached by the stranger a month ago while he was occupying his usual stool at the bar in the Sourdough Saloon. On this occasion he got a few free beers and held a pleasant conversation about the weather. Ciso has a very quick mind and so he purposely neglected to ask his new friend how he had breached our otherwise excellent security perimeter. But he reported back to me the next day, and I realized immediately that some pretty sophisticated technology had been placed at the man's disposal.

On the third occasion, the penny dropped after Ciso relayed to him a little legend we had concocted, intimating that in recent months he had accumulated some serious gambling debts and a nasty drug habit. His friend soon suggested that help might be available, especially if Ciso were in a position to supply in exchange a small amount of local lore. After a short round of negotiation on price, a deal was struck. Then the two of us mapped out a course of information flow and started to feed the demand.

The cash compensation was very rich indeed and the cocaine was of excellent quality. The latter we stashed among the medical supplies destined for our Lake Isabella Palliative Care Facility, where small doses of LSD and cocaine had been found to be most beneficial in the care of end-state cancer patients.

Eventually we put two and two together. "He has to be working for people at Nellis," I remarked one night. "I've told you about the heavy pressures the base commander there has been applying to us recently. He wants open access to our entire territory. This can't be a simple coincidence. Besides, he must have been given military gear in order to be able to bypass our security perimeter so easily."

"Perhaps it's time I confronted him about his provenance, to see how far he'll go," said Ciso. "His latest requests involve the locations giving entry to the interior of the mountain."

So he did, and received in return a frank acknowledgment that the information was being relayed to Nellis Air Force Base, although his contact claimed to be a mere middleman who didn't know anything about why it was wanted. After a high-level conference with Hera and the members of my security advisory committee, we agreed to continue to play along. We also resolved to bring the confrontation with the base commander to a head as soon as possible.

When the final request came it was clear that the endgame had begun. Ciso stalled for time, claiming that it would take at least a few weeks before he could finish getting his hands on such sensitive documents. He also speculated to his contact that he was now likely to be under suspicion at his place of work. Thus he insisted he would obtain what had been asked for only if he could deliver it personally inside the confines of the base. After he handed it over he would want a plane ride to a faraway destination to be named later.

This last deal was easily struck. His handler undoubtedly wondered a bit at Ciso's naïveté in apparently assuming that he would be safe and sound inside the base at the time when no further value could be extracted from his services. Undoubtedly he also attributed this lack of foresight on his part to the presumed effects of regular cocaine use on Ciso's addled brain.

"It's so good to see you again, dear Jacob," Hera said as we entered the small office in the rear of his family's home, located in one of the three Hutterite colonies emplaced around the perimeter of Crater Flat within the Yucca Settlement. "I believe there's a matter you'd like to discuss with me."

"Please sit down. It may be nothing at all, and yet it's something I've become rather uneasy about."

"As I recall, you said it could be a security matter, not right away, but perhaps in the near future. Since that's Marco's area of responsibility, I asked him to join us. What exactly is troubling you?"

"Some time ago I mentioned that over the years I've had occasional communications from the minister of the large evangelical church that's located in the residential compound at Nellis Air Force Base. There have been repeated invitations for all of us to attend one of their services, or alternatively, to have him come here as a guest preacher at one of ours. Lately the requests have become more insistent and have taken on the tenor of commands."

"I don't think I can recall getting a request from you for an exit or entry permit at the security gates for this purpose," I said, "but, of course, you could have one if you like. I think we'd prefer having you go to Nellis, rather than the other way round, but we could assign an escort for a visitor to you, if you wish."

He laughed slightly. "Actually, no thank you. I've discussed this with my congregation, of course, as well as with my fellow ministers at the other colonies. Quite frankly, none of us is interested in the proposition. Before we migrated from South Dakota to take up residence in your territory, first in Jalama Valley and then here, we used to get the occasional visit from the Jehovah's Witnesses. We would always turn them away, politely but firmly, with a mention that if they would simply check one of our websites, they'd discover that our faith has served us well for a very long time, and through some very difficult times. We've never proselytized, and we ask other types of believers to show us similar respect."

Hera interrupted. "I've had my own entreaties along these lines from none other than the base commander at Nellis, General Keith Merden, during the regular reviews of administrative issues arising in our contractual partnership for the nuclear waste facility here. I flatly refuse to even discuss the matter, but lately something of a menacing tone has crept into his requests. The last episode occurred just two weeks ago, when he hinted that we didn't have a legal right to deny entry to what he called the missionaries affiliated with his church. He said then that he expected me to reconsider my position before our next meeting. He wants access both to the town of Beatty and to all our residential compounds in Crater Flat."

"This issue has been building for some time, hasn't it?" I asked. "I've read in some of our intelligence updates that there's a strong evangelical presence throughout the US military and especially in the officer ranks, right up to the generals."

"Yes," she answered. "I asked for those reports to be prepared after I started getting pressured at my meetings with Merden."

"How do you intend to respond?" Jacob asked.

"Marco has a security advisory committee that reports to him. And, as you probably know, our Beatty contingent includes a fair number of retired US military personnel, including some with long experience in strategic planning. We've already run some scenarios in anticipation of some kind of confrontation occurring at some point. Since your colonies form part of our extended family, I'll brief you on them if you like."

"Thank you, but you do remember, I'm sure, that we've always been pacifists, Hera. We will not be taking part in any confrontation. I will pray that you'll be able to resolve the matter without resorting to violence."

"Jacob, I give you my word that I'll do everything in my power to avoid such an outcome, and especially to protect your people from becoming involved, or suffering collateral damage. I will keep you informed."

She turned to me. "You have an urgent meeting with your security committee, I believe. The deadline imposed on us by Nellis is fast approaching. You should go."

I took my leave. Hera had her miniature recording device in place, which picked up the rest of her conversation with Hofer.

"I don't want you to worry about your people's safety, Jacob. My sister Athena is working night and day on seeking to resolve this matter entirely peacefully. The terms of our contract with the US government are crystal-clear on this point: We have unchallengeable authority over the right of entry into the Yucca Settlement, excepting only when war has been declared or when the President has identified a credible terrorist threat at Yucca. Athena has contacted senior lawyers at the Justice Department in Washington who have assured us that our interpretation of the contract is correct. She's also in touch with Homeland Security on a regular basis, and as of yesterday, was told that currently there is no threat of domestic terrorism at our site."

"Then you're confident a confrontation with the military at Nellis can be avoided?"

"Yes, I am. We've offered to send our dispute to legal arbitration, a position supported by Washington, so that should be the end of it."

"Good. Now to other matters. When we were setting up this meeting, you said that, time permitting, you'd like to resume one of our earlier discussions. Why don't we do so now?"

"I'd like that. Some time ago, when I raised the topic of religion's nihilism, you quite properly challenged me to examine my own faith in science. If I recall correctly, you inquired as to how science could avoid a similar charge, in that it gives its practitioners no guidance whatsoever about how they should use the knowledge they gain. Is that a fair summary?"

"It will do as a start. If I may restate my query, I would point to the enormous new powers delivered to humans by science and ask: Do we become better people thereby—in a moral sense? Consider by analogy God's grant to Adam of dominion over all the earth's other living creatures. That was a free gift, unearned by the recipient. Yet it did not vary man's relation to his Creator in the slightest respect. Specifically, it left unaltered the basis of the moral duty of obedience owed to God by Adam and his progeny. It is the Ten Commandments that confirm that the bond uniting man and his Creator is indeed a moral compact. The grant of dominion over the territory of this planet changed nothing in this regard."

"So, you see what science brings as being an enlarged form of this original dominion? I confess I rather like that way of stating the case because the philosopher Francis Bacon, who has influenced me greatly, used such terminology."

"Fine. Moreover, Hera, these vastly enlarged powers are accompanied by what you yourself called a thoroughgoing agnosticism with respect to God. The absence of any inherent moral guidance seems to me to be a good definition of what one ordinarily means by the term nihilism."

"Fair enough, as a starting point. I believe it's accurate to say that the great majority of scientists hold a firm position in this regard. Namely, that science presents new capacities for action and that it's up to societies, through their governments, to decide whether or not to take advantage of them. They don't think it is or should be their own responsibility to make this call."

"I remember your sister Persephone being involved in one of our chats, at the time when I needed to better understand the long debate about the ethical and religious issues surrounding the use of stem cells. I found the information quite fascinating. I had the clear sense that scientists made a serious effort to understand these issues and to participate in those debates, especially with respect to the use of human embryonic tissue."

"Correct. Some of them weighed in directly against the religious viewpoint, suggesting that believers should find a way to overcome their objections in view of the promised health benefits. But most adopted another approach, which in my mind shows more clearly how modern science has evolved in its relation to social norms. The alternative was to say, let's just work around the ethical problem. Let's find other sources of stem cells so we don't have to use embryonic tissue at all. And they did. The controversy evaporated."

"I see. So you're saying that most scientists don't wish to confront such issues directly."

"Many have done so in their roles as citizens and thinkers, but not as something that occurs within the practice of their science itself. The only exception I can think of is a group of atomic physicists who actually thought that the world would have been better off if the discovery of nuclear fission had never been made."

"So you're saying in effect that the charge of nihilism made against science will stick."

She laughed heartily. "Please indulge me a bit longer. Let me try to wiggle out of the box that you're trying to stuff me in. The charge may be laid, but I want to be allowed to examine it more closely before being asked to plead guilt or innocence."

"All right. I will extend the hearing before judging."

"I begin by suggesting that at the very least, science's nihilism is quite different from religion's—although whether or not the difference is important remains to be seen. You will recall from our earlier conversation my contention that religion's nihilism arises in the interface between various faiths, a place I labeled, perhaps a bit overdramatically, I concede, as the dead zone of the spirit. The source of the trouble here is what I call religion's false universalism—the claim that the moral code is universally valid, even though it is regularly violated when believers in other faiths (or other variants of the same faith) are slaughtered in the name of God, such as in the many wars and ongoing conflicts in the name of religion. By its very nature, monotheistic faith is fanatically exclusionary when it comes to either other gods or other conceptions of the very same god. But the conflicting claims simply cancel each other out."

"I recall your exposition on that point. As I replied at that time, from within the perspective of each faith, such as that of my Hutterian Brethren, we worry only about our own salvation. The rest we leave up to God."

"I remember. But not all are so charitable, as you well know. During those troubled times when your faith arose in sixteenth-century Europe, both Catholic and Protestant powers tried to force your communities to convert to one or another of the officially approved faiths on pain of death. Jewish believers, of course, had been getting this charming offer for much longer. But it still goes on. Take the example of the two worst sinners of recent times in this regard—militant Roman Catholicism and militant Islam. Both hold with equal force that each is the one true faith for all humankind. Both remain oblivious to the obvious truth that there are no rational grounds for choosing one over the other. Only the sword can decide the matter, as each well knows from its own history."

"I believe we might have exhausted that theme already."

She laughed again. "Sorry! I do remember that science is the one before the bench in this room today. Believe it or not, I was trying to set up my own plea. By contrast with religion, I regard modern science as the true universal: Everywhere in the world, without exception, those who wish to practice this craft must follow an identical path. But the universal is the method, not the result. The uniqueness of science is that it holds no fixed *a priori* belief about what is or is not the truth about nature. Every result is provisional and subject to further challenge, no matter how numerous its champions may be. Consider the concept of phlogiston in the eighteenth century, or luminiferous aether in the nineteenth. Each of these two alleged substances had a certain plausibility for a time as a decent explanation for observed phenomena. After a while definitive challenges arose against both, made by Lavoisier in the first case and by Einstein in the second, and eventually they were discarded. But there had been no sin attached to upholding these and many other provisional truths."

"As you well know," he replied, "by way of contrast, every aspect of our faith, from the smallest to the greatest, is fixed on a single and unchanging point of belief. We may arrive at our acceptance of God's will by many and diverse routes, which matters not at all so long as we get there. So I see a profound dichotomy here. As you describe it, for science the route—method by another name—is what you worship. Any attained goal along the way is a provisional victory. I get the sense that there may be no such thing as a final truth at the end of this tunnel."

"I suspect that is the case, and not only in some weighty philosophical sense. The standard joke in laboratories is that no announcement of novel scientific findings is complete without a call for additional research."

"Yet worshipping daily before the idol of scientific method cannot possibly yield reliable guidance for how we should behave in life, Hera. Sooner or later the accumulation of new powers will come to resemble nothing more than a magician's act, which depends for its success on distracting the audience from what's really going on, or what's really important to us."

"Just so. Such powers are means enabling us to achieve desirable ends or goals. But as mentioned before, guidance in actually selecting some goals and rejecting others cannot come from within the practice of science itself."

"Then you must rely on some other source external to science for guidance. As you once said, many scientists also remain religious believers, and our discussion explains why."

"That is one solution, to be sure, but not the one I personally prefer. It's just that I don't want to surrender the field so easily. I've not yet given up the idea that an ethical framework may be located precisely where I just said it could not, namely, within scientific practice itself."

"Since you have just contradicted yourself, I must ask how you intend to extricate yourself from the awkwardness of holding two contrary propositions simultaneously."

"Nevertheless I shall try. An ethical standpoint requires that we choose the good over the bad and prefer a greater good over a lesser one. A second, equally essential doctrine requires that each of us takes personal responsibility for our actions and their consequences, in so far as we can reasonably foresee what those consequences may be. I don't think I'll get much disagreement from you on this score."

"Certainly not."

"Most ethical systems, including religious ones, assume or argue that the highest goods are intrinsic, that is, conduct that is good in itself and not because it leads to beneficial results. I hold that seeking to understand both the nature of the universe in which we live, and in what way we ourselves have a share in it, is an intrinsic good. That is so because our minds habitually and naturally are driven by a sense of wonder to seek such knowledge. By the way, I can't take credit for that proposition, which originates with Aristotle. More than any other form of human thought, modern science allows us to pursue that goal, and therefore we can say that it participates in the good, or even that it is in itself intrinsically a good."

"You refer to understanding as an intrinsic good, without precluding the existence of other goods. Do you therefore distinguish it from science's other aspect, previously mentioned—its bestowal of new powers?"

"I suspected you would spot that move, and I do intend that distinction. For what I think are obvious reasons, the operational powers humans derive from scientific knowledge cannot be considered to be an intrinsic good. We can concede that some part of humanity has derived either modest or substantial increases in well being resulting from technological applications of this new knowledge. But I have no idea how one would do the overall net-benefit calculation for all of our race who have lived on earth since this adventure began, say, in the eighteenth century. So far as the rest of nature's creation is concerned, at least, the calculus is far simpler: Those applications have allowed humans to drive a rapidly increasing number of other species into extinction."

"Let me accept what you say for the sake of argument. What follows?"

"In so far as science is driven by a need to understand how the universe as a whole, and our planetary world in particular, came into being, it is, I believe, a good-in-itself. The sheer scope of its explanatory power about matter and energy has no equal in previous epochs of human history. At one and the same time, and using a common method, its gaze encompasses the behavior of everything from immense galaxies in far-distant space to a single-cell organism on earth. And we, too, of course, are subjected to its scrutiny. My position is that this form of knowledge is indispensable to us. We need it in order to know how we should live on this earth."

"I won't comment on its godlessness, for that much is obvious. But surely you can't mean that science is all we need in order for us to know how to behave in this life."

"No. It's necessary but not sufficient. But I don't want to expand on that theme now. Remember that my primary purpose here is only to demonstrate that science may be able to avoid the charge of nihilism. I maintain that such insight about nature as modern science grants us is necessary for us as reasoning creatures who are consumed by wonder. In this respect it is a good-in-itself. You will recall my report of the scientific findings about the origins of the moral sense in primates. However odd it may sound to you, I insist that we humans cannot fully understand ourselves as beings capable of ethical behavior in the absence of such knowledge. Only modern science permits us this insight."

"I have to admire the consistency in your argument if nothing else. And yet the very idea of an ethics that refuses to acknowledge God's dispensation to us is an absurdity in my eyes."

"I fear I have tried your patience, Jacob, and if so, I apologize. Perhaps I should stop."

"No, please forgive my outburst. Really, I meant no censure by my remark. I am capable of fulfilling the role you appointed for me—as a neutral judge before whom you wish to plead. In this role I am supposed to keep my own set of values in abeyance, and I will."

"In any case I'm almost done. My case may be summed us as follows. Close observational studies of our primate relatives allowed us to develop a picture of how moral reasoning evolved in the mammalian brain. The hypothesis is that our far more highly developed systems of values presuppose an evolutionary foundation of this type. Our capacity to form moral minds is originally a gift we received from our mammalian ancestors.

"I'm well aware that you don't need this alternative story, Jacob. But I do, since I can't accept the religious one. You see, before evolutionary biology and neuroscience was far enough advanced to work it out, the secular version of the origins of morality was much the same as the theological one. Both alleged that the values we identify with civilized behavior were superimposed atop a set of savage and unreceptive natural instincts that, once repressed, never stopped seeking a good opportunity to break free again from these artificial constraints. Now we can see how the ground had been prepared long before, allowing a reasoning species to build a larger edifice on these foundations."

"You must still tie the threads of the argument together by explaining more clearly how science, as such, escapes the charge of supporting nihilism. Your alternative story about the natural basis of moral behavior doesn't give science the right to claim the reward for itself."

"The specter that 'everything is permitted' because 'God is dead' has haunted Western societies since the end of the nineteenth century. The clear implication was that an advancing scientific paradigm in evolutionary biology, having destroyed the Biblical account of creation, was incapable of providing a substitute basis for ethical systems. There seemed to be no escape from the bleak moral vacuum where only the raw power of dominant interests or dominant predators ruled. But now we know better. The idea that there are evolutionary origins for moral minds provides robust support for all human efforts to strengthen the values that protect human dignity—and, increasingly for many humans, that are seen to protect the dignity of other species of animals as well.

"I will concede immediately, however, that science may claim this prized status only if it is willing to pay the required price."

"And that is?"

"To renounce and retract its gift of powers. Not necessarily forever, but certainly for the time being, while anarchy prevails in international relations and everything is permitted. This act of renunciation can only be performed by communities of scientists themselves—ultimately, all of them working together in concert—because all other social groups are far too eager to exploit whatever operational powers they can get their hands on."

"But didn't you say earlier that most scientists don't see how they could or should be held responsible for what is done by others with their discoveries?"

"For some the very idea of accepting personal responsibility for the outcomes of their work is deeply offensive. They refuse to acknowledge that as a result of their attitude, science had entered its own form of the moral vacuum. To me, they're the modern world's version of Pontius Pilate, who turned Jesus over to the lynch mob. Isn't it obvious that the great powers flowing out of science could not exist without their work? Only scientists launch their discoveries into the social world. Only they have both the unambiguous moral authority and the duty to withdraw them again."

"I could see the logic if you said they had the capacity to do so, since they are its creators in the first place. But why do you say *moral* authority?"

"Because it follows from my earlier point, from my claim that the knowledge they generate is a good-in-itself. Its inherent goodness can be protected only if the knowledge is rigorously separated from its applications. There is a specific reason for this stipulation because one core question lies at the heart of every experiment: How does nature work? Modern science's approach to its subject matter carries within it an inherent predisposition to action. Upon its successful conclusion, we know how to duplicate some natural process whenever we like—and also how to apply it in other circumstances that might be most beneficial to us. But from this characteristic, it follows, I maintain, that an ethical duty arises, namely, each scientist's duty to examine the range of potential applications in the context of existing conditions."

"Are you saying that such an examination must be made for each and every experiment or observation that has ever been done? Even I can see how this might be impractical."

"And you would be perfectly justified in saying so. No, I believe that the duty to which I just referred is something that arose at a particular point in time, namely at the onset of the Second World War. The community of atomic physicists faced the agonizing prospect of imagining atomic bombs in the hands of Hitler and his allies. A revered member of their fraternity, Albert Einstein, when faced with the horrors of the First World War, had likened modern technological progress to 'an axe in the hand of a pathological criminal.' He had been expressing himself metaphorically. Then, a mere twenty or so years later, his metaphor had become all too real.

"So, what I'm saying is that around 1940 the enterprise of modern science as a whole, around the globe, had arrived at a turning point. Any doubts that it had done so should have been erased a decade later when the nuclear arms race between capitalist and communist states was in full swing. By 1955 the hydrogen bomb was being deployed; after about another decade, there were intercontinental ballistic missiles with multiple, independently targeted warheads. Within a mere quarter-century from the moment of the first realization, in 1940, there was enough destructive power just sitting and waiting for the launch signal to put an end to human civilization, perhaps forever."

"I think I see the full argument now. The moral duty you mentioned applies to science as a whole, and thus, of course, to the community of scientists as a collectivity, and not to any specific experiment. But I am still a bit confused by the references to moral authority as well."

"That is correct. It's the collectivity to which I refer. By 1962, say, at the time of the Cuban missile crisis, it was obvious that things had gotten completely out of hand. That is the moment when senior scientists might have begun the process of gradually withdrawing their discoveries from the public realm."

"Again, is this not a similarly impractical idea? Wouldn't it be easier for them to abandon entirely the practice of doing science?"

"Well, as it happens, Jacob, two of the last century's most distinguished physicists, Max Born and Albert

Einstein, expressed precisely that sentiment toward the end of their lives. But I don't agree because of my firm conviction about the intrinsic value of scientific knowledge. I think the new discoveries should be withdrawn from public circulation indefinitely, and thus from both military and commercial application, but that the enterprise of discovery must go on."

"I would have reiterated my objection about impracticality, except that I now recall what you've told me on prior occasions about the large university operation you've set up in Las Vegas. You are running it according to the principles you have articulated here, are you not?"

"Yes, but not by ourselves. We're part of a small global network of similar bodies. You see, it may well have been entirely impractical for the community of scientists to do this in the 1960s, although we have to avoid claiming that this couldn't possibly have happened just because it didn't. The general collapse of industrialized societies that began a few decades ago gave us our chance, and we took it."

"The part about moral duty is clear, at least in your own terms. You don't want Einstein's axe thrown into the fire, but you do want those who are discovering ways of making bigger and better axes to keep their findings concealed for the time being. Now I want to ask you one more time about moral authority."

"If scientific knowledge is an intrinsic good, Jacob, then the discoverers and possessors of that good are the bearers of a sacred trust. They have the right and obligation to set the terms under which the intrinsic good they have brought into being may be converted into something entirely different, namely, utilities or useful things. Therefore they must first assess the types of powers that may be developed from their discoveries. Second, they must assess the capacities of governments of the day, around the world, to act together to prevent those powers from being applied to horrible ends. They may seek assistance from others in making such assessments, but they cannot avoid taking personal responsibility, as members of a global collectivity, for the likely consequences flowing from what they discover."

She fell silent, and after a few moments had passed, Jacob said, "I will have to attend to other business soon, but I wanted to say something else before we part. Some time ago I learned from Marco about your sister Io's death. I have not had a chance to express my sympathy and condolences, and I wish to do so on this occasion."

"Thank you, Jacob. We sisters have been unusually close throughout our lives. This has been a very difficult time for me because over the years I assumed responsibility for making a number of decisions that affected Io's life. I am still experiencing terrible bouts of self-doubt and guilt as a result. I cannot seem to be able to forgive myself. And as you see, I cannot even speak of it without beginning to weep."

His voice was so gentle as to be nearly inaudible. "I know you cannot accept what I am about to tell you, but I am compelled to say it anyway. We can seek forgiveness only from God. Only He can lighten such a burden for us. Nevertheless, I pray that someday you will find your own path to inner peace. I suspect you have sought succor in your music."

"Your words are a great comfort to me. Yes, I have listened over and over to all of the great requiems in classical music. And to some of my other favorites as well, where I made a wonderful discovery. May I relate it to you? It helps me to talk."

"Of course."

"I listened over and over to Beethoven's late quartets, and especially to his opus 132, the work in A minor. This contains a long third movement that the composer entitled 'a holy song of thanksgiving to the divinity by a convalescent.' The year was 1825 and Beethoven had been gravely ill for some months, so much so that he himself fully expected to die and all his friends thought likewise. But miraculously he recovered his health; thus at the outset of this movement, he marks the score 'sensing new strength.' Then at the beginning of the closing adagio we find the expression, 'with the most fervent and intimate feeling.' So, what do we expect to hear as this extended movement draws to a close? One would think that such a song of thanksgiving, celebrating a narrow escape from death's clutches, would conclude by bursting out in notes of wild, delirious joy. But no, not here. The tone is, rather, one of infinite, almost unbearable sorrow. How are we to understand this extraordinary paradox?

"To be sure, the validity of the answer to this question that welled up within me as I listened is purely personal. For what it's worth, here it is. Beethoven had indeed been at death's door, and in his mind he had watched this dread portal recede again as he regained his health and creative spirit. This was his just reward, perhaps—I'm speaking metaphorically here, of course—for the immense suffering he had endured in his life."

She spoke so softly, Jacob could hardly hear her.

"What suffering?" he asked.

"By middle age he had gone completely deaf. Can you think of a more punishing trauma for a composer? And if that was not bad enough, he also endured a severe form of tinnitus, described as a continuous roaring in his ears. One can only guess at the unending agony this caused. But in addition to these and many other bodily afflictions, I believe he also suffered greatly in a spiritual sense, as a result of forcing himself to extend his musical ideas into those realms of the beyond where most of us cannot follow. There he glimpsed the eternal, and the sadness this revelation engendered in him is to be understood thus: As the authentic response of a mortal mind to the realization of the necessity of its own demise. This is the only way I can explain for myself why a song of thanksgiving would be expressed in terms of such unrelenting sorrow. Beethoven tells us that lingering forever in the realm of the eternal is denied to us. We can envision this realm in our minds while we live, but we cannot dwell there after death."

"For most humans the realm of the eternal is understood as God's kingdom," he said.

"Not for Beethoven, who chose his words carefully here. The German word he uses to denote the addressee of his music is *Gottheit,* which refers not to the God you worship, Jacob, but to an abstract idea of such a being. The word translates into English as deity or divinity, as far as I know. Which serves as a useful reminder that one can be religious in many different senses."

"Indeed I regard you as a religious person in the same way as you claim Beethoven was, Hera. Clearly your mind also senses the reality of the transcendental realm that lies beyond the bounds of our mortal life. It is as if, together with the rest of us, you approach this sacred realm with fear and trembling and then turn away at the very last moment, just at the point where the Kingdom of God beckons."

Now finally she allowed herself to laugh again. "I am so very grateful that you never give up on me! Sometimes I think I was corrupted by those writers who turned the Biblical stories about God and the fallen angels into poetic dramas. I think especially of Milton and Goethe."

"How did they corrupt you, if I may ask?"

"By giving Lucifer by far the best lines."

"Ciso was called into Nellis Air Force Base by his handler a few hours ago. They were expecting him to bring the remaining sets of tunnel diagrams and entrance codes for our sites, and as far as I know he has done so."

Our little war council was assembled in a command bunker buried deep beneath the foot of Yucca Mountain, and I relayed to the group what I had just learned. We had been meeting daily for the past couple of weeks, but this meeting had been called only an hour ago.

"Then I expect we'll get our own summons shortly," Hera said to me. It was just past 5 AM.

With me were Hera, Gaia, Athena, and three senior operations personnel from our staff support base in Beatty. The latter were worried about their families.

"We won't let the town of Beatty become a battleground, I promise you," Hera told them. "As soon as we're ordered to do so, we'll draw back the security gates and declare it an open city. In fact, we don't expect to see combat operations anywhere in our territory, but, of course, we can't be certain of that. We'd thought of bringing all of the Beatty residents into the Yucca Mountain tunnels where our people and the Hutterites will be assembling soon, since there's plenty of room. But the military drones overhead are watching every move we make, and they'd spot the traffic in an instant. Try not to worry; whatever happens, your families will be kept out of harm's way."

I faced the group. "I also need to tell you that al-Dini and his wife and children were picked up by a jeep from Nellis around midnight last night outside the university grounds in Vegas. They were taken to the base. We had him under observation and therefore we know he had the briefcase with him. We're also able to track it electronically and the signal says it's somewhere in the admin building on the base. We would know if they'd succeeded in opening it, and therefore we can conclude that they haven't."

The phone rang and I answered. "Tell him we'll leave by copter in ten minutes. We should be at the rendezvous point in half an hour." After disconnecting I gave them the news we were all expecting. "General Merden has requested an urgent meeting with Hera and me. The rest of you have your assignments, I believe."

As we traveled toward the city, she remarked, "I think we need to let our own plan play out. We have no assurance that our friends from Washington will arrive in time, or even if they do, that Merden won't keep them away from us until it's too late."

"We have to assume that he's ready to go—in fact, for all we know he's already given the orders for his troops to move against us. I think he wants to present the Washington contingent with a *fait accompli.*"

"How convenient it is for us now that years ago we were barred from ever entering the Nellis base for unspecified security reasons. Our regular meeting place is about a mile west of the base perimeter, isn't it?"

"Yes, a little over a mile, I think." I had my binoculars trained on the base's residential compound as we banked to turn nearby, and I spotted Ciso's truck parked in the vicinity.

"Is the wind direction as expected?"

"Pretty much as usual; westerly."

There was a beep in the cockpit. "We've reached the beacon," I said. "The signal must be sent. It's now or never."

"Do we have any choice?"

"None that I can think of."

"Then send it."

Although we couldn't directly observe the events that were scheduled to follow in the next millisecond, we were reasonably sure they had transpired, since we had run the experiment successfully at our own test facility many times. A moment before, somewhere in the bowels of Nellis's central administration building, in the room where al-Dini's dark metal briefcase had been deposited, a group of technicians almost certainly had been still puzzling over how to open it. Dr. al-Dini himself was probably in attendance and experiencing directly the exasperation of his new colleagues. The access codes for all of the buildings, offices, and labs throughout our entire university complex were contained inside, as he had confirmed to his satisfaction when inspecting it the previous day, and the others were anxious to extract the information.

But, of course, it was not the briefcase al-Dini thought it was when he had hastily assembled his family and a few possessions before his flight. To anyone who had only glanced quickly at it, the case we had substituted was identical in every way. The present one, however, was fashioned out of the very latest nanometal compounds, and when his handlers had first greedily taken possession of it, they failed to spot the highly unusual feature that later caught their attention: Nowhere on the entire surface of the small cubic mass that lay before their eyes was the smallest seam detectable. Nor were there any latches, just a handle affixed to one side.

But the radio signal broadcast from our helicopter as it passed nearby solved their puzzle. Small explosive charges inlaid in the metal neatly blew apart the device along a series of concealed seams. A second set of charges detonated immediately thereafter and a mist of tiny particles invisible to the naked eye filled the room and began its longer journey through the building's ventilation system and into all of its many rooms.

At exactly the same instant, Ciso's huge pickup truck appeared to disintegrate of its own accord. The side panels along its entire length were ejected some distance away from the vehicle. High-pressure spray nozzles embedded in the chassis were activated and, aided by a light breeze, a much larger invisible cloud of a similar material dissipated and began to settle over the entire residential area located inside the gated perimeter of the base.

No sooner had our helicopter touched down than our escort party approached and we were led into the low building where all of our liaison meetings had been held. As we were frog-marched down the corridor, Hera murmured to me, "I think we both know the name of the person we're very much missing at this instant." She didn't have to mention Phil's name.

As we entered the room, General Merden neither greeted us nor extended his hand. He fit the mold—tall, lantern-jawed, thin-lipped, blue-eyed, beribboned, and immaculately suited in a fresh uniform. His junior officers remained at some remove from their leader.

Hera's face was an unreadable mask. She challenged him as soon as we had walked through the doorway. "Why did you cut our satellite communications link during the night?"

"We have identified an immediate and credible terrorist threat at the nuclear waste facility." He glanced at his watch. "It's just gone 6 AM. You have two hours to open all access gates around the perimeter of the territory you have illegally sealed off. My troops have orders to enter this area at precisely 8 AM today, whether or not you have complied."

"What exactly is the nature of this so-called terrorist threat?" I demanded.

"That is classified information. There is nothing else for us to discuss. My adjutant will escort you back to your copter."

"Just a minute. You have been notified that senior officials from Washington are right now on their way out here to settle this dispute between us about access. So please forgive me for saying that under the circumstances, the imminence of this threat seems a bit contrived."

"Just about two hours from now my troops will have found what we'll be looking for inside your fences. We plan to arrange a display of that evidence in time for our visitors to have a look at it soon after they arrive."

"Something tells me that the evidence is already in your possession and has never left the base," I suggested.

"Get them out of here!" he yelled to his minions.

Hera was quicker. "You're quite correct, you know. There is a terrorist threat in the vicinity. But you're mistaken about the location. It's actually on your land, not ours. And it's not a threat anymore. It's already happened."

At a glance from the General, one of his underlings leapt to the nearest phone and asked to be connected to base headquarters. He wasn't smiling when he hung up and turned back to face his commanding officer.

"HQ has just confirmed two small explosions on the base about 15 minutes ago. One was outside, near the residential compound. The other seems to have gone off somewhere in the basement level of the admin building. No casualties have been reported."

"Two small explosions? No casualties? That's all?" the General barked.

When Hera next spoke, her voice was so controlled and even that for a moment I thought I was attending a diplomatic reception. "Unless I am very much mistaken, whoever was in the building at the time has already begun to die. The same goes for your wives and children."

Absolute silence ensued. The junior officers looked to their leader, who simply sneered.

"What a crock of bullshit. How would you know what's happening on my base?"

Then the light dawned. "Unless you two are the fucking terrorists." Whereupon he drew his sidearm and his men followed suit.

"Shoot me if you like," she replied. "They won't die right away, or even in the next day or two. But mark my words: They will surely die soon thereafter unless they receive the proper medical treatment. And I'm the only one who knows where the medicine is."

He holstered his weapon again. "You're just bluffing to gain time until your friends from Washington arrive. Exactly what are my people supposed to die from?"

"Massive pulmonary hemorrhaging due to inhalation of anthrax spores."

"Bullshit. Anthrax is treatable. That's a well-known terrorist threat and we've got a good stockpile of antibiotics right on the base." He turned to his adjutant. "Get the lowdown on this situation. Now! If there's a suspicion of anthrax, get the emergency response protocol activated."

She held up her hand. "No antibiotic in the world that you have stockpiled will be of the slightest use in this case."

"And why not?"

"Because the anthrax strain now contaminating your base has been genetically engineered to be resistant to every known form of antibiotics. You'll be wasting your time in administering whatever drugs you have. All you will accomplish will be to prolong the agonies of the dying."

"You've just confirmed that you are the terrorists." He looked again at his officers. "Handcuff them both." Then back to us: "You can either tell me right now where the medication you're holding is, or I'll send some troops to blast their way into your facilities and find it."

We already had our arms pinned behind our backs when she retorted. "Stop being such a damn fool! Would we have come to this meeting if we were planning to use a bioweapon against you? I know what material was used because last night it was stolen from one of our secure labs. And it was brought onto the base by two men who worked for us and who were recruited into your service by agents acting on your own orders."

Once again his officers looked to him for new instructions. Then the cuffs came off and as they did so, Hera motioned in the direction of the conference table. "There's absolutely no risk of immediate death to anyone on the base—although I must caution you, the infection has begun its work. Shall we sit down and review how to respond to the situation sensibly?"

The hatred visible in his eyes fought a brief struggle against his better judgment and lost. He moved to the table, saying as he went, "If you're conning me about this, I'll have your head on a platter for it."

She appeared unfazed but I could detect some signs of the strain she was under. "And I warn you in return that our stores of medications are housed in a secure facility and protected by a booby-trapped door. If you send your troops in with guns blazing, there'll be nothing left to treat your people with. You'll have your proof about what I'm saying, and sooner rather than later. Now, can we cut the crap? I don't respond any better to threats than you do. What other questions do you have?"

This time the hatred won. "Just come clean and tell me if you did this."

She let him stew a bit before responding. "You appear not to be listening, so I'll repeat what I said. There are two men from our territory who are right now on your base and who were escorted there by your own people. I'm quite sure that you know what I'm talking about. One is named Jesús Narciso Ramirez, the other is Abdullah al-Dini. Dr. al-Dini is a professor of biology at our university, and I believe his wife and children were brought with him. The other gentleman, Mr. Ramirez, is—or was—our assistant chief of security. Would you now please confirm the whereabouts of each of them?"

The General glanced at his men, but it was already clear that he had been fully briefed. "They are both presently on the grounds of the base. The family is there as well. What do they have to do with the anthrax attack? And, if I may ask, what the hell are you doing with this stuff in your labs?"

"As to your first question, they both vanished suddenly last night, and we only found out about it early this morning. The anthrax went missing at the same time. There is no other explanation. The only real mystery is how and why they ended up in your hands."

"Answer my other question first. The only possible reason why a private party such as you would have weaponized anthrax in your possession is to carry out a terrorist attack. So?"

"The answer involves classified information, but under the circumstances I'll overlook the usual restrictions. For more than twenty years we have had a joint-venture operation in which the other party is none other than the US government. If you doubt my word, you can get all the confirmation you need as soon as the party from Washington gets here. Our business involves research and experiments on engineered pathogens that could be used as bioweapons, and, of course, on countermeasures. Now it's your turn. What are those two doing in your hands?"

A junior officer approached and whispered in his ear. He nodded and looked up. "The plane from Washington is a half-hour away, but they can't land as scheduled at Nellis since the whole base is locked down. They'll have to use the runway you maintain at the old Las Vegas airport."

"I'll make the call," I said. "They'll be expected. But none of your vehicles that are on the base should be used for the pickup. Do you want us to make the arrangements to have them brought here?"

"No. We have plenty of vehicles parked outside, even though there are more in the party than expected. Apparently my superior officer is among them."

Hera decided to exploit her advantage. "We knew that as of yesterday, even though you didn't, General. You have underestimated us rather badly, I'm afraid. Surely you must have been aware that one of your recent predecessors in this command, General Philip Ziegler, was my partner up until the time he was murdered. We still have very good contacts in the upper echelons of the US Air Force."

For the first time since the meeting began, he appeared to have nothing to say, so Hera resumed. "Marco and I don't see any point in continuing our discussion with you until the other officials get here. What we propose to do is to return to our helicopter and call the people who are manning our own emergency response facility. We have a large and well-trained biohazard team. We'll set in motion the necessary steps. I'd like to have the phone number for the officer in charge of biohazards at Nellis; we'll need to coordinate with him or her."

Without raising his eyes from the table he simply nodded at his officials who then escorted us back to the tarmac. Once inside, a very relieved Hera collapsed into my arms while our pilot and co-pilot began to place the calls. When Athena and Gaia picked up, she moved to the console at once. "Everything's going to be OK. Spread the word, especially to the Beatty folks. Sound the all clear so that people can leave the tunnels. And get the biohazard team ready to go. We'll get back to you again shortly."

The first person she greeted warmly when we re-entered the meeting room was our good friend General Gary Hone who, as base commander at Vandenberg Air Force Base, had helped us immeasurably in the terrible period following Phil's death. Now second-in-command of the US Air Force and based in Washington, he was everything in appearance that Merden was not, being rather short, portly, and avuncular. After she finished hugging him he burst out, "Why didn't you just call me once you started having problems with Merden?"

"I didn't want to bother you with something that seemed so trivial. I really thought the issue was going to be resolved easily."

"Well, it has been resolved; he's been arrested and will be court-martialed." His look quickly turned serious. "How big a problem do we have at the base? Are we looking at hundreds of very sick women and kids as well as soldiers?"

"Call the others over and sit down." When the Washington group and Merden's junior officers were assembled around the conference table, she resumed. "The only area of the base that's been contaminated is the admin building. The material that was sprayed out of the truck near the residential compound is a completely harmless powder. We use the apparatus built into this vehicle for testing how well a type of bioweapon material has been aerosolized and thus how well it disperses under various weather conditions."

"Thank God for that!" He turned to the officers: "How many soldiers were in the admin building at the time of the explosion?"

"About two hundred, sir," came the reply.

"There, too, the situation isn't as bad as it could have been," she told him. "The sample taken from our lab has a relatively low anthrax spore count in the silica mixture, making it much less dangerous than some of the other preparations we have on hand. Still, it is a bioengineered version and we've created the only antibiotic that will stop the infection."

Hone paused. Several lines of thought seemed to be working simultaneously through his mind before he asked, "Has the drug been dispatched to the base?"

Hera nodded. "Our biohazard unit is on its way and will join up with the similar military group that's part of the Nellis complement. They know each other—they've carried out a number of joint exercises. They'll enter the building through its loading dock and set up a personnel decontamination facility inside. After people go through the facility they can leave the building. They then report to a medical team that will start all of them on the antibiotics regimen our people will be bringing. Everyone who shows signs of infection will be monitored continuously until the danger passes. When treatment is started this early, a one hundred percent recovery rate is virtually guaranteed."

"What about the building and its immediate surroundings?"

"The biohazard people said that the doorways were sealed off right after the explosion, and the ventilation system was shut down, so it's unlikely that very much material escaped. As far as the interior is concerned, however, there will be so much contamination that the best thing to do is to tightly seal all the apertures in the structure permanently and just leave it that way. Sorry."

"So be it. We'd already decided to close Nellis for good and shift its operations elsewhere. In fact, Merden had been informed of this decision months ago. We really don't need three big air force bases in the southwest anymore. But to be frank, there was also a lot of unease in Washington about the fervor of the evangelical spirit that Merden has been cultivating there for a long time. I'm pretty sure that his little anti-terrorism caper was designed to show us how much we needed his Christian warriors."

I had waited for a chance to intervene. "There are two of our people on the base. I need your authorization to go pick up one of them. His name is Ramirez and there's no reason for you to detain him. He wasn't involved in the anthrax attack and the officers here say he hasn't been anywhere near the admin building. But he was bribed by agents working for Merden to steal copies of our site diagrams. We'd like to deal with him ourselves if it's OK with you."

He turned to the officer group. "Two of you go with him to the base. Find this Ramirez and turn him over to Marco." We left at once.

"Hera, what about the other one Marco mentioned?"

"That would be Professor Abdullah al-Dini, and he's reported to be somewhere inside the admin building. He's a renowned specialist in molecular biology and was recently assigned to our bioweapons section in Las Vegas. As far as we know, he stole the anthrax and is the one who released it. Alternatively, of course, it might have been released accidentally. We do know that he, too, was recruited by Merden's agents; he probably told them about the anthrax and they decided to add it to their arsenal. Anyway, I imagine you'll want to arrest him as soon as he's been through the detox procedure and leaves the building. We'll be as interested as you undoubtedly are in the results of your interrogation of him and his handlers. He'll probably deny all knowledge of the anthrax, of course, but eventually you'll probably get the truth out of him. And you must promise me there will be no torture."

He laughed. "You have my word on it. And you know as well as I do that only the entertainment industry still thinks that's a good way to get reliable information."

She grinned gently as she rose to shake his hand. "I should get back to check on the clean-up operation. We can meet again tomorrow morning for a progress report. By the way, the whole group from Washington is under my orders to come to Beatty for an evening before you go back. We have a small banquet facility attached to the Sourdough Saloon that's run by a first-rate chef."

I ran into her at the helicopter pad just as I arrived back from the base with Ciso in tow. She looked at him and they both grinned. "When we get done with you, Mr. Ramirez, you'll wish you had remained in military custody." We all had a good laugh. "Seriously, now, we will never forget what you did for us. I will say more about it at the big celebration dinner I'll be throwing."

When we were alone for a moment after returning to Yucca, she said, "Hone is a dear, dear friend and we owe him a great deal. I feel quite badly about deceiving him."

"He's a very sharp fellow, mother. It won't take him long to figure out that there may be more to the story than he's been given so far. But one of the first things he'll discover is that Merden was indeed just hours away from launching a full-scale assault on our territory this morning, and that the terrorist threat was a pure invention on his part. He may have plenty of suspicions about the anthrax attack, but he's also a very busy man, and as long as everyone is restored to health, he'll probably switch his attention to some of the many other problems he has on his plate."

"I realize I gave a pretty lame excuse for not contacting him earlier, but I didn't want to confess the real reason.

There won't always be someone on the other end of the telephone call to come to our rescue when we face serious trouble. We needed to get some experience in handling this kind of crisis all by ourselves."

She fell silent for a moment and then looked back at me. "What do you think will happen to al-Dini?"

"Hard to say. They might let him go after a while if he tells them that he must have taken the wrong briefcase by mistake and that he didn't even know where our bioweapons labs are located, which is true enough. It may be the military officers who were running him who wind up in the biggest trouble—after all, they brought the briefcase onto the base without examining it first. But I wouldn't be at all surprised if one of them tries to cut a deal by trying to convince the interrogators that they had heard about the anthrax from al-Dini and had asked him to procure it. They can say they didn't know about our joint-venture work on bioweapons with the government and that they feared an attack on the base. If I'm right, then al-Dini's own fervent protestations of innocence will fall on deaf ears." I paused. "Are you feeling sorry for him?"

"No. We offered him and his family a safe haven and the thanks we got was an act of betrayal. He'll have plenty of time for repentance now. But I do want you to find his family and offer them a chance to return to live at Vegas. They're blameless and we have plenty of job openings to fill."

One day several months later I arrived at Hera's suite to find her relaxing with Gaia and Athena. "We need you to go to Washington, Marco."

"Is there some residual unhappiness over the anthrax episode? Have I been selected to take the fall for our group and to be the counterpart in a prisoner swap for al-Dini?"

They all laughed. "Probably not, at least not as far as we're aware," Hera replied. "The government partners in our bioweapons joint venture have had a quiet word with their colleagues in the air force. The whole thing will be treated as an unfortunate accident and the file has been closed. They've accepted our offer to carry out the procedures for sealing the building. And it seems Abdullah may be released soon, although he won't be returning to our area, apparently at his insistence."

Athena added, "The base has been evacuated and most of the installations and military equipment has been moved to Edwards and Vandenberg. The Nellis site will be turned over to us, and responsibility for providing air cover for Yucca Mountain has been transferred to the Edwards base. You'll be pleased to know that they've acceded to your request to leave behind a small fleet of the surveillance drones and their ground control facility, as well as a few military transport planes. They'll train some of our people to operate the drones."

"Excellent," I declared at once. "We can abandon what's left of the Vegas airport and use Nellis instead. It's huge; it was once the largest air force base in the United States. Their controllers sat at video consoles and launched missile strikes from unseen Predator drones cruising all over the globe."

"That whole enterprise has been transferred to Vandenberg. The officers and their families are delighted to be moving to the coast. After the city, the casinos, and the golf courses all shut down here, there wasn't a lot for them to do except hide from the heat."

"Obviously you all know what this means. We'll be all alone here. There'll be no one else for hundreds of miles except some wandering tribes in the desert here and there. Which reminds me, we'll need to take possession of the secure rail corridor from Vegas to Bakersfield and the coast. We can't ship our people and the supplies we need all by air."

"That's one of the topics on the table for discussion when you get to Washington. They like the Bakersfield facility a lot, since our ongoing blood surveillance program guarantees them early warning of changes in infectious pathogens, which is one of the things that worries people the most. They want us to set up a couple of similar satellite facilities—one in San Diego for a start—although on a much smaller scale. There aren't many people left in the San Diego area, but it's still on a migration route along the coast from the south, and they're anxious to monitor the types of pathogens folks may be bringing with them from Central and South America."

"What are we being offered in return? And will there be enough food? What about electricity and water for the new hospice?"

"The number of Hutterite colonies in the area around Vandenberg has been increasing because they continue to subdivide, so the extra food won't be a problem. The federal government has offered to restore and secure the coastal rail line running to San Diego, and to supply a small desalinization plant and another one of the compact nuclear reactors to support the new satellite operation. Both of the plants will be installed in La Jolla because what we get in return is the right to occupy the whole area where the Salk Institute, the University of California campus, and the Scripps Institution of Oceanography used to be. Fortunately those sites were secured when they were abandoned and there wasn't much vandalism. If it all works out, we'll transfer some of our university operations, except for the bioweapons part, from Las Vegas to La Jolla.

"Here is a briefing book for your Washington meetings," Athena concluded, "that includes drafts of the new contractual agreements based on our earlier models. The schedule is on top. And we'd like you to leave tomorrow."

"If you've got an hour or so to spare right now," Hera added, "why don't you stay? Before you arrived the three of us were reviewing our recollections of the little conversations we held with Abdullah before things went sour. We were trying to figure out why we failed so miserably in trying to persuade him that we had the best interests of science at heart. You were there for all of them; maybe you can help us figure it out."

"Maybe you're looking for complicated reasons when there's a simple explanation at hand," I replied. "He fled from a region where the governments and religious authorities were keeping a close eye on what scientists were doing, at least in his own field. Maybe what he found when he arrived here sounded uncomfortably familiar to him, despite, what are to you, the obvious differences between your objectives and the ones he had confronted earlier."

"You may be right, although I still find the whole business with him rather upsetting," Gaia remarked. "Whatever the ultimate source of al-Dini's problem with our agenda, I'm concerned that we didn't articulate our rationale more clearly in those sessions with him. One of the main reasons we got together today is to finish what we started."

"I confess I have another agenda as well," Hera noted, "although I agree with Gaia about the need for greater clarity in how we deal with scientific research at our university. We've never had another case as serious as al-Dini's in our recruiting to date, although perhaps we've just been lucky so far. My more personal objective is to resurrect a thought that first occurred to me many years ago, when I fantasized about being able to bestow the name *Homo scientificus,* or more precisely *Homo sapiens scientificus,* on our tribe. If we agreed that we wanted to do so, what kind of claim would we be making? Wouldn't we have to demonstrate that we had diverged so sharply from our contemporaries and competitors, whom I prefer to label *Homo religiosus,* that we're entitled to have a distinctive appellation on the evolutionary tree?"

Athena chuckled. "You've smuggled a most subversive idea into your question, as I'm sure you're well aware, sister. Like all of its predecessors along the evolutionary line of placental mammals leading up to the appearance of hominins, the genus *Homo* evolved through a number of stages, all but one of which turned out to be dead ends. I'm thinking of earlier species such as *H. habilis, H. erectus, H. ergaster,* and *H. heidelbergensis.* None of them chose their traits or their capacities, for those were shaped by natural selection; and they certainly didn't pick their own name! Your quite naughty idea is that with the emergence of *H. sapiens* and its reflective self-awareness, the possibility arises that a species of *Homo* might actually create its evolutionary significance as a product of its own free will."

Gaia interrupted. "Before you two start riffing on this wild theme, may I make a more prosaic point? There may be a thematic connection between my own practical issue, about the demand for us to explain our objectives as managers of a scientific research establishment, and Hera's desire to pinpoint the 'species-being' of *H. scientificus.* What they have in common is the need to articulate why scientific practice has a special meaning for our lives."

"I would go further," Athena added, "and say that we see modern science as a watershed in human development, a decisive change in course or direction for humankind."

"We could push your metaphor a little harder, Tina," Gaia remarked, "and regard the coming of Bacon's new sciences as the terminus of the childhood of humankind. For the first time, they allow us, once and for all, to leave behind the terrifying domain of good and evil spirits and replace it with a testable explanation of why things happen the way they do."

"And I can name the person from whom you appropriated that idea, sister," Hera said to her. "Its source is one of the passages from Condorcet's wonderful book, his sketch of *The Progress of the Human Mind,* which you included in the little memo you prepared for the first of our three conversations with Professor al-Dini. Do you remember? I can quote the passage from memory because his mode of expression is so striking there. He claims that the method of analysis worked out in the new sciences 'has for ever imposed a barrier between mankind and the errors of its infancy.' He and the other thinkers of the period known as the Enlightenment believed that the main battle had been won. They were certain that the change of course in human affairs had already been made."

"I do recall those words," she answered. "He thought that in his own age, at the end of the eighteenth century, his fellow humans were ready and able to 'put away childish things,' if I may quote the Apostle Paul here without giving offense. But his timing was a bit off, wouldn't you agree? The time was not yet ripe for the triumph of the scientific ethos in social relations."

"Very far from ripe, as subsequent events proved," Athena commented, "but in most instances an excess of optimism is a forgivable flaw. He was only fifty-one years old when he took his own life while imprisoned and awaiting certain execution. Had he lived another twenty years he would have witnessed the Bourbon Restoration in 1814. The counter-revolution that followed brought to an end any further progress in realizing the Enlightenment dream for many decades to come."

"Yes, in retrospect, we can see how his mind had leapt ahead and anticipated much that would eventually come to pass in social changes—although not for almost another two centuries," Hera added, "and then only in parts of the globe. There was another kind of anticipation entirely in his thinking, for it took almost as long for the natural sciences to emerge within society as a distinctive force in their own right. Only toward the end of the twentieth century did it become crystal-clear that many saw the progress of those sciences as an indispensable ingredient in humanity's hopes for the future of the species."

"So is that the answer you've been looking for?" I asked. "*Homo scientificus* begins to emerge when the new sciences are recognized as a vital part of society's agenda? Is that the time when Condorcet's ghost would have breathed a sigh of relief and exclaimed, 'Now, at last the long-awaited moment has arrived'?"

"No, most certainly not! That's the mistake I tried and failed to get al-Dini to understand. By the time the moment you mention had arrived, scientists had already surrendered their autonomy. You remember, that was the point made in the discussion about Szilárd and the other atomic physicists in the 1930s and 1940s. A few of them glimpsed briefly the possibility that they might take control of the scientific enterprise and set the terms under which society could use their discoveries. Then this vision vanished again during the Cold War.

"A few decades ago some molecular biologists experienced a similar epiphany, but they also failed; it was a case of too little, too late. They had finally realized why their technologies of gene manipulation posed dangers greater even than atomic bombs—because the techniques they had perfected were so much more subtle, at least in terms of outcomes. It was the exact opposite of the mighty explosions that could rip whole cities to shreds and make them uninhabitable due to radioactivity. For with many genetic alterations, especially in brain function, there wasn't even any direct evidence of what had been intended. All one saw, sometimes much later, was the behavioral outcomes, and who could say, in the absence of a definitive test, whether or not the resulting oddities had actually been planned by someone? Besides, the techniques themselves had become so simple that anyone with a modest amount of training and equipment could have a go at it."

"You have exposed effortlessly the logical flaws in my remark," I said sheepishly, thus generating considerable merriment around the table. "Let me attempt to restate the case. You appear to be implying that the emergence of *Homo scientificus* does not take place merely because scientists found out how to hand over to their patrons a set of operational powers that turned humans into the masters of the universe. Correct, so far?"

"Correct, because as the saying goes, power corrupts," Gaia replied. "What we heard from al-Dini was the well-rehearsed mantra of the scientist who just wants to get on with the process of discovery. Frankly, I find the lack of any sense of personal responsibility for what might result from his discoveries simply astonishing."

"Don't be too hard on him," Hera said. "Like almost everyone else, he has no sense of history. He's never reflected on where he stands in relation to the past and future of his sciences. I admit, you did try to convey a sense of the larger trajectory with your collection of stories about some leading scientists. But I don't think he ever really connected on an emotional level with those biographies."

"Well, sister, we tried. You helped me make the point about how things started to fall apart as Europe stood on the brink of the First World War. On the side of social relations, there was a clear sense then that Condorcet's Enlightenment project was still alive after all, having survived underground during the long reign of the nineteenth-century conservative reaction to the French Revolution. It was the Industrial Revolution that re-kindled it. Around the turn of the twentieth century there were the beginnings of a social security system, workers' rights, and, of course, the political enfranchisement of women."

"That was one side of the picture we drew. The other concerned the continuing progress of the natural sciences. The Western world was entering the period when science would emerge into the public realm as something quite special. The signature event in this regard happened shortly after the end of the First World War. In the early 1920s, the persona of Albert Einstein suddenly leapt into the popular imagination through the press. It began with the British expeditions to Brazil and West Africa in 1919 to take photographs during an eclipse of the sun, a project designed to seek confirmation of the prediction in Einstein's relativity theory about the bending of light rays. When the astronomer Arthur Eddington announced that this empirical test of the theory had been passed, what was amazing was how the story exploded on the front pages of the world's newspapers. In the years that followed, when Einstein and his wife made extended tours to the United States and Japan, he was treated like royalty, indeed better. Not only was his itinerary a matter of public record, but reporters badgered him with questions on all kinds of social and political issues. The age of the scientific sage had arrived."

"And then just as quickly it all started to fall apart, didn't it?" Gaia remarked. "At exactly this same moment, the rabble-rousing by Hitler and his little gang of thugs in the streets of Munich had already begun. Fast forward a mere ten years later and you see the same Einstein fleeing for his life, his achievements mocked in the land of his birth by a regime that referred to relativity theory as 'Jewish physics.' After the full truth of the Holocaust emerged, he resolved never to set foot on German soil again."

"So," Hera added, "in pulling these threads together, we say that about a century after Condorcet died, that is, around the year 1900, Europe had finally arrived at the crossroads for the Enlightenment project. The two sides of the project—the sciences of nature and the sciences of society—stood ready to join forces at last, because the most important aspect of Condorcet's vision was that each side would support the other, in a kind of reciprocal action. That prospect had thrilled him. What he could not foresee, of course, was what might happen if the reciprocity collapsed."

"Of all the intellectual giants who contemplated the ensuing wreckage," Athena interjected, "only two were utterly shattered by it—Einstein and his soulmate, Max Born. Their correspondence gives us a sense of the depth of the bitterness and regret they experienced. For them the great adventure of science was over, finished for good. For it was obvious that the world wasn't really interested in the intellectual adventure of science. What it had wanted from science all along was power."

"Still, it was an offhand comment by Born in one of his letters to Einstein, written just months before the latter's death, that offers a clue to an alternative solution," Gaia noted. "Modern science should carry on, but do so in private, just as it had done when it first arose. But then, at the point of origin, it was an accident of circumstances. The early proponents of the new sciences first had to give the world a little peek at the riches still locked inside nature's mysterious structures before industry and the state were ready to acquire the enterprise. And so they did. But in our time, now that the Enlightenment project has been rejected by most of humanity, there's an explicit choice to be made. The new sciences must be hidden away again because it's irresponsible and self-defeating to place the powers they offer in the hands of those who long for the End Times."

"What you just said is a truth of the twenty-first century that Born and Einstein could not have anticipated," Hera replied. "Sometimes my musings about *Homo religiosus* strike others as a bit quirky, but really, what more does one have to do other than glance at the world around us? We have to conclude, once and for all, that the Enlightenment thinkers who saw religion as a feature of humankind's infancy were simply wrong. There is a profound need for a story that explains in a few easy lessons how the person is related to the cosmos. There is an equally profound desire to believe in the persistence of some aspect of the person after death. And there you have it. By way of contrast, science is very hard to grasp and always points to what is not yet known, and it has nothing at all to say about an afterlife. It can't compete."

"True enough," Athena added. "But maybe more time is all that's required before what you call the Enlightenment project can triumph. Maybe it was never realistic to think that religion could be so readily displaced by science. Remember, there's evidence of religious worship in the Kalahari Desert—the deity was a snake, if I recall correctly—dated seventy thousand years ago."

"I don't know whether you've just made your point or mine," Hera answered. "Religion's great antiquity, combined with an utter lack of any plausible evidence for its truth, which hasn't improved one iota in those seventy thousand years, tells me I'm dealing with a belief structure that's impervious to influence."

"Yet, you would have to concede that your precious modern sciences arose from exactly the same brain and mind, in the same species," she retorted.

"Touché. I therefore leave open both possibilities—one, that religion will continue to collect the support of most people, as far in time as one can see into the future; the other, that it will gradually fade away and be replaced by science. Whether I choose one or the other, nothing changes in terms of where our responsibilities lie, and they, too, are twofold. First, we must protect the future of science. Second, we must hide it away so that the powers it reveals can no longer be deployed in the world of men. At least not for the time being. Our successors can only uncloak it again if and when the conditions for realizing Condorcet's dream of reciprocity have ripened at last."

"I fully agree," Gaia said. "And, fortunately, we don't have to presume to hoist such weighty responsibilities onto our shoulders alone. Now that our domestic crisis has passed, I can embark upon my long-planned tour. As you all know, a little network of like-minded institutes has sprung up here and there around the globe, and I want to visit them and also invite them to pay us a call here in return. If we're to have the slightest hope of succeeding in our quite mad endeavor, we must coordinate our plans and activities."

"I think I recall you telling me that all of them are co-located with nuclear waste facilities. What's the final list of destinations?" Athena asked.

"Three in Europe—Sweden, Britain, and the joint Franco–German facility at Gorleben; also Canada and Japan. At one time or another, all of them had the same idea we did: A well-constructed deep geological repository for storing nuclear waste and designed to last for thousands, if not tens of thousands, of years is an excellent place to hide other important things as well. Of course, there's plenty of high-level nuclear waste in many other countries in the world, but none of them got their act together before the social chaos struck."

"Have all of the other five added scientific research facilities and support structures for staff and their families within their security perimeters?"

"Yes, to a greater or lesser degree. Perhaps none is as elaborate as ours, except the ones in Germany and Japan. One of the reasons for my trip is to discuss whether or not we should each specialize in a few fields of science. My guess is that this will happen. But all of the five partners also believe that we have here a superior level of both resources and security, due to our location in the desert, to be able to host the only comprehensive facility. I expect to sign a memorandum of agreement to this effect during my visits. If the plan goes ahead, we'll need to get ready to receive a fair number of staff and families who'll be transferred from the other institutes."

"Too bad there's not a facility in Italy on your list," Hera muttered, "because in that case I'd join you. I'd love to go back there. I still remember my grand tour, even though it occurred something like twenty years ago."

Athena chuckled. "We're all well aware of your opinion that the northern Italians may justly claim to be in possession of the finest culture and cuisine in the known universe, sister. However, even you might concede that they've had their problems with governance over the years. I'm pretty sure that no one in Europe was keen on the idea of storing high-level nuclear waste anywhere within easy walking distance of the nearest branch office of the Mafia. The consensus was to ship the Italian waste to Germany."

"Speaking of the Mafia, I presume that as you go traipsing around the world, you'll be accompanied by Marco and a security detail?" Hera asked.

"That's already arranged, as is military transport, courtesy of General Hone," I reassured her. "We can leave right after I've concluded my business in Washington. Narciso can look after security matters while I'm away."

"There's much to do in the meantime, so I'd like to draw this little colloquy to a close," Gaia announced. "The bottom line for me is that with this alliance, we're well on our way to achieving our practical objective. Our partners are in full agreement with us on restricting the flow of scientific research results to our internal communications. We're setting up a joint peer review process for publications and grant awards. Our MOU will also include a collaborative evaluation for all proposed transfers of intellectual property to external parties. Basically, we want to lock down any innovations that could be easily weaponized. So if a bunch of fanatics wants to slaughter their neighbors in the name of the one true God, they'll have to do it with axes and spears rather than with dirty bombs or engineered pathogens."

"What about the other agenda item, Hera?" Athena inquired. "Did we make any progress in helping you to define the nature of *Homo scientificus*?"

"Enough for now, because in the end I think my question may not be all that different from what Gaia just referred to as our practical objective. In other words, a group of humans may be entitled to bestow that name upon themselves only under certain conditions. Above all, they must be able to demonstrate that they know how they must care for that precious body of knowledge, entrusted to them by past generations, which is like no other form of knowledge ever revealed. I refer to the sciences of nature, with their unbreakable unity between profound insight and access to unimaginable powers."

"In order to exercise this type of care," I ventured, hoping to redeem myself from my earlier embarrassment, "they must set in place the rules articulated by Gaia a moment ago. And those who are authorized by the community of scientists to administer these rules on their behalf will constitute the priesthood of science."

She gave me a beatific smile. "So they must, or devise some other scheme that will be as good or better in reaching the desired objective. Your formulation is felicitous because nothing is more certain than that there will be no viable solution without such authorization. The basic idea is excruciatingly simple: Those who make the discoveries must accept responsibility for the uses to which they may be put. The scientists who have taken up residence in the six institutes do so because the obstacle preventing this acceptance in the past has been removed. For until now there was no mechanism in place to overcome it, no way for them to escape the Hobson's choice of either disclaiming personal responsibility or giving up the practice of science altogether. The terror of being forced to steer between a rock and a hard place is gone."

"I recall your relating to me what was said in your conversations with Jacob Hofer about nihilism in religion and science," Athena commented. "The procedures now in place at Solomon's House and its affiliated institutes mandate the severing of discovery and application. Consideration of practical uses involves an entirely separate process in which matters of ethics, risk, and social forces are uppermost. It seems that what we have done is to rescue science from the prospect of falling any further into its own black hole of nihilism. Thus science becomes once again the master of its own fate."

"The modern sciences are a little bit like religions in this respect," Gaia added, "in that they must not serve any master other than their own inner guiding principles. They must exert their autonomy vis-à-vis the powers that be in society, whether political or economic. This is none other than the mode of operation of Solomon's House. How remarkable it is that this truth should have been divined so clearly, so long ago, by Francis Bacon!"

"Your words were well chosen just then," Athena said, "but don't forget it was originally this same thinker whose grave error it was to encourage his followers to think of themselves as the prospective overlords of nature. This they can never be, for how is it possible to imagine that one of nature's own creations, however clever and resourceful, could set itself up in the ruler's chair? One massive asteroid set on a collision course with the earth would dispel that conceit once and for all."

"I'm always searching for an apt metaphor to convey the meaning of an abstract idea," Hera remarked. "For me nature is the sorcerer, humanity the apprentice. The disciple needs to be wise enough to realize that she can't always be entirely sure she has learned all that is necessary in order to make the magic spells do only what they're expected to. Francis Bacon would have enjoyed Goethe's story, I think."

"We may have different preferences in metaphors, Hera, but I'm pretty sure we agree on what we take away from them. When science once again knows no master other than itself, and thus is driven solely by the search for understanding, then, by virtue of this state of self-control, it can legitimately claim to have arrived at mastery. Not mastery over nature, but mastery over its possession of the keys to the powers of nature."

"I like that formulation, sister. Now, are we done?" Gaia asked.

"Just about," Hera answered. "But first I have a dream to relate to you."

The mention of this additional contribution caused a series of exaggerated groans to reverberate around the room.

"It's short, I promise you. I dreamt I was standing alongside Frodo at the lip of the abyss inside the mountain of Orodruin in Mordor. Gollum was hot on our trail, of course. Frodo—who funnily enough looked a lot like Marco—reached over for the object I was carrying, and to my surprise it was not the ring of power, bearing its elegant Elvish script, but rather a small box. The sides and cover of the box were all inscribed with a beautifully rendered scientific notation; I'm not sure, but I think the symbols might have been from Einstein's equations of general relativity. Inside was a set of computer storage drives containing the sole remaining copy of all the publications from the history of modern science. As Frodo looked into my eyes I saw the terror in his own gaze when he screamed, "Give me the box, we must destroy it *now.*"

She paused and I couldn't resist taking the bait. "What happened?"

"I woke up, thank goodness."

Note to the Reader

Go to www.herasaga.com

Study Guide and Index, including brief explanations of technical terminology, references, and web-links;

Maps and Pictures, containing photographs and web-links illustrating the locations referred to in the text;

Imaginary Futures, a short essay and bibliography on the tradition of utopian literature in modern Western thought, which begins with Thomas More's *Utopia* (1516), especially the fascination with science and technology in that tradition;

Further Reading, a short bibliography dealing with themes in the text;

Scientific Articles related to themes in the text.

See also "About *The Herasaga*" at the end of this volume.

🖰 = web link available at www.herasaga.com

Further to the Dedication

My mother, born in Brooklyn, New York, in 1911 and named Ethel Bertha, was the daughter of a German butcher and his wife. She did not finish high school and went to work as a corporate secretary in New York City at the onset of the Great Depression. She married my father, William Leiss, Sr., a house painter, in 1933; I was the first of five sons. In 1947, now with four boys aged seven and under, she followed my father to rural northeastern Pennsylvania, to a 50-acre forested property five miles outside the small town of Honesdale. There were no other houses visible from ours, which was without central heating, an indoor toilet, or running hot water. The well dried up in summertime, so she hauled the laundry down to the nearby brook. We bathed in an aluminum tub set in front of the kitchen wood-and-coal-fired stove.

The two oldest boys were enrolled in a one-room schoolhouse—one teacher, twenty pupils, grades one through eight—some miles away. With the brook, the lake that was part of our property, and the endlessly exciting stretches of woodlands, it was an idyllic place for young boys to grow up, especially in summertime. My mother would have used other words to describe her fate, but as a *Hausfrau* of her times she was loyal to my father's quixotic dream. (She had been heard to mutter that, if she ever slipped on winter ice while carrying the shit bucket to the outhouse, the experiment would have ended then and there.) One failed income-generating scheme succeeded another as my father tried to supplement the meager wages he brought home after a 50-hour workweek. But we had large gardens and the land was rich in wild berries; the fruits and vegetables my mother canned throughout the summer and fall months helped us through the long winters.

During our first five years in the wilderness, my father had a deep well drilled and installed a coal-fired furnace and steam radiators, an indoor bathroom and septic field, and a propane-fired kitchen appliance to replace the cookstove.

The fifth boy was born early in 1953; later that same year my father, while painting the third-story façade of a department store, fell to his death from a scaffold onto the sidewalk of the town's main street. I was thirteen. Paradoxically, it seemed to her, the workman's compensation and social security checks provided us with a cash income slightly higher than what his hard-earned wages had totaled.

Many years later my mother told me that in her first years as a widow she had had a recurring nightmare, a story in which it seems he had only been injured in his fall and somehow vanished in the aftermath, reappearing at home months later—whereupon she was forced to hide him on our property so as not to have to forego the greater financial security his death had occasioned.

She had been a regular but indifferent churchgoer all her life, and every Sunday we boys were dutifully hauled off, freshly scrubbed and dressed in our one good outfit, to a nearby tiny Protestant Church at a crossroads named Indian Orchard. (The building still stands in front of the cemetery where she and my father are buried.) But in the first anxiety-laden years after my father's death, she turned to the Bible for support, reading it again and again from cover to cover, which led to a serious rupture between us when I announced my loss of faith upon returning from my first year at university.

This rift tormented her, and in order to resolve it she began rifling through the boxes of course books I left behind, including scholarly works on Ancient Near Eastern religions. The epiphany she soon experienced hit her like a thunderbolt, and thereafter she could hardly restrain her sense of outrage over the fraud that, she felt, had been perpetrated on her in the name of faith. I can still recall the look of resolve on her face, while she watched the local Jehovah's Witnesses innocently walk up the long driveway to our house, as she marshaled in her mind the arguments—mined from her personal collection of the Bible's less savory tales—she would soon hurl in their astonished faces.

She was forty-two when her husband died and lived without remarrying for another forty-four years. During the first two decades of her long widowhood, she was consumed by a single goal—to hold us together as a family while her children got the early schooling that would assist our passage out of rural poverty.

Acknowledgments

During a trip to Nova Scotia in the summer of 2003, my partner, Jeanne Inch, introduced me to her old friends Alex and Rhoda Colville and their daughter, Anne Kitz. Jeanne had first met the family as a child growing up in Sackville, New Brunswick, where her father, Robert Inch, was a colleague of the painter at Mount Allison University. Subsequently, the artist gave me permission to use three of his paintings as cover artwork for this trilogy. I am most grateful to Alex Colville for this gracious act, and to Anne Kitz for assisting me with the arrangements for securing the transparencies and permissions from the museums that hold two of these works.

Church and Horse (1964) has been reproduced with kind permission from AC Fine Art Ltd; *Horse and Train* (1954) is used with gracious permission from the Museum of Fine Arts in Montreal; *Moon and Cow* (1963) is reproduced courtesy of the Art Gallery of Hamilton.

In at least some respects the painting style known as "magic realism" dovetails nicely with the minor literary genre of utopian fiction. Both ask their audiences to accept their obvious violations of normal experience as the point of entry into another dimension of existence. More particularly, both seek to achieve their effect on us in the same way, namely, by concealing beneath the work's surreal surface layer an elaborate, precisely drawn architecture—in the painting, an exact geometry of space, and in the fictional work, a methodical dialogue about ideas. The intended effect is, of course, a conjurer's trick, the creation of an illusion. By willingly suspending disbelief, the audience can pass through the work's portal and live for awhile in another dimension of space and time.

The short piece entitled "About the Herasaga," at the end of this Back Section, contains an interpretation of the imagery in the three Colville paintings used as cover artwork in this trilogy. No one should regard this brief interpretive exercise as anything other than my own purely idiosyncratic reading of these settings. I imply no claim that the painter himself had any such meanings in mind for his creations.

Jeanne's ongoing tolerance of a writer's obsessions for a little project that began innocently enough in the summer of 2002 is truly remarkable. Holly Mitchell has been a loyal supporter of it from the outset, as has the little readers' circle organized by Mary Smith of Kingston. But the schedule for Book Two has largely been determined by the insistence of Holly's mother, Phyllis, now eighty-seven and a lively resident of an Ottawa retirement home, that the empty slot reserved for it on her bookshelves, next to her copy of *Hera*, should have been filled some time ago. I have relied on Richard Smith's generous help and advice on many aspects of this enterprise, and other old friends—Steve Kline, Ian Angus, Mike and Kathy Mehta, Scarlett Schiaperelli, Steve Hrudey, Harrie Vredenburg, Cooper Langford, Éric Darier—provided encouragement when it was most needed.

Marc Saner of Ottawa, trained in both biology and ethics, gave me a careful reading and a detailed commentary on the draft that was extremely helpful. I owe a debt of gratitude to Ed Levy of Vancouver, who did a close critical reading at my request both for this book and its much longer predecessor, and to Julia Levy for her perceptive comments on this one. My dear friend and comrade-in-arms Gilles Paquet adopted my orphaned trilogy at the University of Ottawa Press and entrusted this manuscript to the care of its capable staff, Marie Clausén, Jessica Clark, and Eric Nelson. As they had done for *Hera*, the dynamic Winnipeg duo of Jenny Gates (copyediting) and Brian Hydesmith (production and cover design) turned it into a handsome and readable volume.

The McLaughlin Centre at the University of Ottawa, which rescued me from mandatory retirement some years ago, is recognized worldwide for the high quality of its academic work. There its director, Dan Krewski, and his assistant, Suzanne Therien, not only provide a comfortable base for my professional work, but also cheerfully indulge me in my more whimsical enterprises.

Sources

The sources for the four epigraphs that follow the Table of Contents are as follows:

Francis Bacon's *New Atlantis* (1627): A work of merely some forty pages, it is breathtaking in the power of its anticipation of the future. It is still in print in many different editions and is also available in its entirety on the Internet.

Albert Einstein, Letter to Heinrich Zangger, December 6, 1917: *The Collected Papers of Albert Einstein*, vol. 8, *The Berlin Years: Correspondence, 1914-1918*, English translation by A. M. Hentschel (Princeton University Press, 1998), p. 412.

Max Born, *The Born-Einstein Letters,* translated by Irene Born (New York: Walker & Company, 1971): letters dated November 28, 1954 (pp. 229-230) and January 29, 1955 (pp. 232-233).

For Einstein's life, see Jürgen Neffe, *Einstein: A Biography*, translated by Shelley Frisch (New York: Farrar, Straus

& Giroux, 2007). His writings on society and politics have been collected in *Einstein on Politics: His private thoughts and public stands on nationalism, Zionism, war, peace, and the bomb*, edited by David E. Rowe and Robert Schulmann (Princeton University Press, 2007). Also useful is Thomas Levenson, *Einstein in Berlin* (New York: Bantam, 2004).

An indispensable source for tracking Leó Szilárd's struggle to contain the threat of nuclear fission is *Leo Szilard, his version of the facts: Selected recollections and correspondence*, edited by Spencer R. Weart and Gertrud Weiss Szilard (MIT Press, 1978). Weart's *Scientists in Power* (Harvard University Press, 1979) tells the story of Frédéric and Irène Joliot-Curie, who discovered artificial radioactivity in Paris in 1934 and shared the 1935 Nobel Prize in Chemistry. By 1939 they and their colleagues had worked out the theory of a chain reaction and a design for both a nuclear reactor and an atomic bomb using uranium—which explains why Szilárd was so worried about what was happening in Paris!

In early February 1939, Szilárd, then in New York, learned that the French team was preparing to publish a paper suggesting that bombarding uranium with neutrons would lead to a chain reaction. He sent Joliot-Curie a letter urging him not to publish, but not knowing who Szilárd was, the French physicist thought that this completely unexpected request was bizarre. After some further transatlantic exchanges, Joliot-Curie made up his mind and sent the following telegram: "QUESTION STUDIED MY OPINION NOW IS TO PUBLISH REGARDS"; the paper appeared in *Nature* in April 1939. (For a full discussion, see Weart and Szilard, *Leo Szilard*, chapter 2, and Weart, *Scientists in Power*, chapter 5.) After the Nazi invasion of France in May 1940, the Joliot-Curies entrusted some of their research notes and materials, along with a quantity of heavy water that had been recently purchased from Norsk Hydro in Norway, to two young colleagues, Hans von Halban and Lew Kowarski, who proceeded to smuggle them into England. Joliot deposited the rest of their research notes in a vault at the Académie des Sciences building in Paris just before the Nazi occupation of that city, where they remained hidden for the duration of the war.

The French team, as well as American, British, German and Russian scientists, had all concluded early on that heavy water would be the best moderator for a nuclear pile, but all also knew that this substance was difficult to produce in sufficient quantity. So most of them also tried to work with graphite; this material, however, was hard to obtain from industrial sources at the very high level of purity that was required. In 1940 both French and German research teams made errors in their calculations of how well graphite would serve as a moderator, leading them to abandon this line of inquiry. Early in that same year, Szilárd and Fermi, working at Columbia University, did the calculation correctly, showing that a graphite-moderated reactor could work. Szilárd sent their paper on this subject to the journal *Physical Review* and asked that it be accepted but not printed. When he told his co-author Fermi about the hold on publication he had arranged, Fermi exploded, saying this was "absurd"; only the intervention by their boss at Columbia, George Pegram, who supported Szilárd, kept the paper from being published (Weart, pp. 144-145). Thereafter Szilárd worked like a man possessed to find a source for the graphite of requisite purity. At the University of Chicago in early December 1942, they used graphite blocks in the world's first controlled nuclear fission reaction.

The correct graphite calculation was the first of three decisive nuclear discoveries—and three terrifying secrets—from that fateful year, 1940. The second followed in short order:

> In late May Louis Turner, a physicist at Princeton, sent Szilárd a copy of a paper showing theoretically that when uranium-238 absorbed neutrons ... a new element, plutonium [would be produced]. Although Turner did not realize it, he had written the prescription for the easiest route to building a nuclear bomb. Szilárd wrote back at once to say that a paper of his own was being kept secret, which implied that an official move was underway to withhold papers. He persuaded Turner to write the *Physical Review* and delay publication. It was well he did so, for Turner's paper could have been an essential clue for the Germans and others. (Weart, p. 145)

The third secret discovery was made in Britain by Otto Frisch and Rudolf Peierls. Frisch was a physicist and Austrian Jew who had fled to Britain upon Hitler's accession to power in 1933. He traveled frequently back to the Continent, especially to work with Bohr in Copenhagen. He was Lise Meitner's nephew, and he visited his aunt shortly after Otto Hahn's 1938 paper (on the collision of a neutron with the uranium nucleus) had appeared. Hahn, a chemist at the University of Berlin, could not explain the result he had achieved; it was Frisch and Meitner who realized that atomic fission had occurred, and whose resulting paper on the subject contained the first public use of that term. Peierls was a German physicist of Jewish ancestry who had also fled in 1933. Their joint 1940 secret paper, known as the Frisch–Peierls Memorandum, showed that only a kilogram of pure uranium-235 was needed to sustain a chain reaction, and that if such a reaction was uncontrolled, it would produce an explosion of staggering magnitude. (This calculation is the one that

Heisenberg supposedly did incorrectly, thus greatly overestimating the amount of U-235 needed to make a bomb; a good discussion of the controversy on this point is available online .)

Three fateful discoveries from the year 1940 that, if they had become known to German scientists, might have prompted the Nazi regime to embark upon an atom bomb project. A single individual, Leó Szilárd, appears to be solely responsible for the fact that two of the three remained secret. One may not unreasonably refer to all this as a close call.

§ § §

Max Born (1882-1970) came from the Prussian city of Breslau, which was also Fritz Haber's birthplace. Although the families of both of them were part of the Jewish community in that city, it is hard to imagine any two individuals more different in character than were Haber and Born. Again like Haber, he met Einstein at the University of Berlin, where Born had been appointed professor of physics in 1915, before his move first to Frankfurt and then, in 1919, to Göttingen.

In one of the narrative sections of *The Born-Einstein Letters* (pp. 149-151), Born relates a charming story from the period of social chaos in Berlin after the collapse of the German government at the end of the First World War. Workers' and soldiers' councils were being set up everywhere during the winter of 1918-1919, and university students followed suit. At the University of Berlin one day, they kidnapped the Rector and other senior officials and marched them to the Reichstag, where the student council held its meetings.

Already in 1918, Einstein was a famous public figure as a result of his opposition to the war. So someone asked him to try to act as a mediator, whereupon he telephoned Born and proposed that he and a third colleague should go to the Reichstag together. There, in a public meeting, Einstein tried to persuade the students that academic freedom should be protected at the university. The students listened politely, but declined to set the officials free, so the three emissaries walked over to the Reich Chancellor's office to seek a meeting with President Ebert. There, too, Einstein was recognized at once and the three of them were ushered into the president's office.

Nothing came of their mission that day, but Born comments: "We left the Chancellor's palace in high spirits, feeling that we had taken part in a historical event and hoping to have seen the last of the Prussian arrogance, the Junkers, and the reign of the aristocracy, of cliques of civil servants and the military, now that German democracy had won." In April 1933, however, his name was on the list of civil servants dismissed from their posts by the Nazis because they were Jews; he eventually found refuge at the University of Edinburgh. The depth of the personal bond forged between Born and Einstein during their Berlin years is evident on every page of their correspondence, but they never saw each other again after 1932. Their common feelings about science and society, illustrated in the epigraphs that appear after the Table of Contents, survived even Einstein's profound unease with quantum mechanics, for Born was one of the great founders of that discipline.

Gaia's memo in Chapter 8 tries to give a sense of the extraordinary community of geniuses who flocked to Göttingen in the 1920s under the patronage of Max Born and James Franck. Among the young masters who were sent there to study with Born was Werner Heisenberg. The two of them, along with Pascual Jordan, were nominated for a Nobel Prize by Albert Einstein in 1928, but when the prize for the discovery of quantum mechanics was finally awarded in 1932, it went to Heisenberg alone. Born was bitter about this, of course. Right after the announcement, Heisenberg wrote privately to Born that the award was unjust since it should have been made to the three of them jointly, and later, in the 1950s, he said so publicly.

Born received his own Nobel only in 1954, the same year Linus Pauling won his first prize, this one for chemistry. (The second, the Nobel Peace Prize, came in 1962; in the early 1950s, Pauling came within a hair's-breadth of beating out Watson and Crick in the race to discover the structure of the DNA molecule, and had he done so, without a doubt he would have garnered another Nobel.) Born's comment (p. 231) is: "I cannot say with any certainty whether I was right that the simultaneous award of the Nobel Prize to Linus Pauling and myself had anything to do with the fact that neither of us had anything to do with the practical application or the misuse of science for political purposes." Although he had not been recruited into the war effort, Born supported the participation of those who were, including in the American atom bomb project: "For under the given circumstances nothing else can be done to save the rest of our civilization." After he returned to Göttingen upon his retirement, Born (along with Heisenberg, Max von Laue, Wolfgang Pauli, Otto Hahn, C. F. von Weizsäcker, and others) was among the signatories of the "Declaration of the Eighteen from Göttingen" (1957), which opposed the re-armament of West Germany with atomic weapons.

In a letter to Einstein dated July 15, 1944, Born noted that a newspaper in Scotland had reported "that you have called upon intellectual workers to unite and organize some protection against new wars of aggression and to secure their influence in the political field. I was very glad when I read that" (p. 144). In this same letter Born advocated the formulation of "an international code of behavior or ethics" for scientists and named Bohr in this context. A few months later Einstein replied (p. 148): "It is, of course, quite correct for you to allot the relevant priesthood to Neils Bohr."

The quintet of Bohr, Born, Einstein, Pauling and Szilárd will always have a special place of honor in that section of the pantheon of science that is reserved for those who were tormented by the role of science in society.

Heisenberg is a special case, of course, due to the never-resolved ambiguity (attributable to Heisenberg himself) about what was said during his notorious visit to Copenhagen, in occupied Denmark in 1941, to meet with his mentor, Neils Bohr, and then again on his return visit to Bohr in 1947, accompanied by a British army officer, because at that time Heisenberg was "in custody" in the hands of the Allies. The initial meeting is the subject of the well-known play by Michael Frayn, *Copenhagen* (1998); the Anchor Books edition (New York, 2000) has a useful essay by the play's author on the lingering controversy.

The critical point about all this is what role Heisenberg had played in the wartime German evaluation of the possibility of building an atomic weapon. Thomas Powers wrote a long, fully researched and carefully reasoned book on this subject, *Heisenberg's War: The secret history of the German bomb* (New York: Little, Brown & Co., 1994), which is an indispensable resource for anyone interested in this subject. Powers concludes that Heisenberg made no effort during the war to try to persuade the German authorities that making an atomic bomb was feasible. In a footnote (note 27 on page 508) Powers writes: "The efforts [in the United States] of Arthur Compton and Ernest O.

Lawrence in the fall of 1941 to convince Vannevar Bush that a bomb was feasible are recounted in detail in Rhodes, *The Making of the Atomic Bomb*, pp. 347ff.... On a theoretical level, the German and American research programs were at this point probably neck and neck. If Heisenberg had shared Compton's zeal, a genuine race for the bomb probably would have followed, lasting until war's end."

The unique combination of events, during which a remarkable collection of individuals was creating a fateful new science at just the time when Europe was falling into the grip of fascist barbarism, continues to inspire new commentaries. Among the most interesting of recent works is the book by the University of Pennsylvania physicist, Gino Segre, *Faust in Copenhagen: A struggle for the soul of physics* (New York: Viking, 2007). His father, Emilio Segrè, who won a Nobel Prize in Physics in 1959, was a student of Fermi's in Italy in the late 1920s. When Mussolini passed a law in 1938 banning Jews from holding university positions, Segrè just happened to be on a summer visit to the University of California at Berkeley. He managed to find refuge there and later worked on the atom bomb project in Los Alamos—another one among the flood of talented Jewish scientists, all highly motivated by the fear that the Nazi regime might beat them in the quest, who were no longer in Europe where they might have been pressed into wartime service by the fascist regimes as a result of the murderous anti-Semitism that raged there.

Faust in Copenhagen tells the story of the last in an annual series of meetings involving seven physicists arranged by Neils Bohr in Copenhagen. The year was 1932, and present in addition to Bohr were Max Delbrück, Paul Dirac, Heisenberg, Paul Ehrenfest (a close friend of Einstein's), and one woman, Lise Meitner. (A seventh invitee, Wolfgang Pauli, decided to skip the meeting.) After their scientific discussions had been concluded, they staged a skit, written primarily by Delbrück, in the form of a parody of Goethe's *Faust* (1932 was the hundredth anniversary of Goethe's death).

In the skit, Bohr was cast as the Lord, the absent Pauli as Mephistopheles, and Ehrenfast as Faust. (Ehrenfast, who had succeeded H. A. Lorentz at the University of Leiden, suffered from severe depression and in 1933 he killed himself and his son, who was afflicted with Down's syndrome.) As Segre explains, 1932 was a decisive year for atomic physics, with the discovery of both the positron and the neutron—the latter ushering in the idea of nuclear disintegration, the first step on the road to nuclear fission. And decisive for European politics as well, of course. Of the seven participants in the series, all but Dirac and Bohr were Germans or Austrians. Four of them were of Jewish or part-Jewish ancestry (Bohr, Ehrenfest, Meitner, and Pauli). Only Dirac and Heisenberg survived the ensuing catastrophe living in their native countries.

§ § §

Some of the material in the Prologue refers indirectly to a mode of thinking that goes by the name of "evo-devo," short for "evolutionary developmental biology." There are many useful articles on this subject on the Internet, including the Wikipedia entry. For a longer treatment, see *Genes in Development: Re-reading the molecular paradigm,* edited by E. M. Neumann-Held and C. Rehmann-Sutter (Duke University Press, 2006).

The source for the epigraph for Part One is Dr. Jane Rogers, who was then Head of Sequencing at the Wellcome Trust Sanger Institute, Cambridge, UK. She was quoted by BBC News on December 4, 2002, at the time when the journal *Nature* had published a special issue on the sequencing of the complete mouse genome, which the journal has made available in its entirety, free of charge, on the Internet. (The mouse genome sequence itself is also available.) This is one example of the remarkable "public face" of molecular biology.

A 39-part made-for-television series entitled "ReGenesis" explores science-in-society issues with specific reference to genomics, and at its own website, the Ontario Genomics Institute offers a very useful "episode guide" with further discussion of those issues. The guide for Episode 9 deals with the public availability of at least part of the smallpox virus genome, which attracted worldwide attention resulting from the success of a reporter for *The Guardian* (UK), James Randerson, in obtaining a fragment of smallpox DNA through the mail (June 14, 2006).

Those interested in Narciso's story about his biological origins, related in Chapter 1, may wish to read the article by Jamie Shreeve, "The Other Stem-Cell Debate," *The New York Times,* April 10, 2005, and in particular, his discussion about the research program of Rockefeller University's Dr. Avi Brivanlou. Another researcher commented on Brivanlou's program as follows: "'Literally nobody wants to see an experiment where two mice that have eggs and sperm of human origin have the opportunity to mate and produce human offspring,' says Dr. Norman Fost, professor of pediatrics and director of the bioethics program at the University of Wisconsin and a member of the National Academy of Sciences committee reviewing stem-cell research policies. 'That's beyond anybody's wildest nightmare.'"

Further to the discussion toward the end of chapter 2, see Marc Hauser, *Moral Minds: How nature designed our universal sense of right and wrong* (New York: HarperCollins, 2006) and Frans de Waal et al., *Primates and Philosophers: How morality evolved* (Princeton University Press, 2006). In the de Waal collection, the essay by Christine Korsgaard is especially relevant, e.g., the sentence on p. 118: "The distinctiveness of human action is as much a source of our capacity for evil as it is of our capacity for good."

For more on the issue of traumatic victims' memories and military applications, see Jonathan D. Moreno, *Mind Wars: Brain research and national defense* (New York: Dana Press, 2006), pp. 128-132.

The source for the epigraph for Part Two is *The Physicists,* a play by the Swiss writer Friedrich Dürrenmatt originally published in 1962. The English translation is by James Kirkup (New York: Grove Press, 1964), and the passage is on p. 76.

On the life and career of Fritz Haber (Chapter 5), one can consult Daniel Charles, *Master Mind: The rise and fall of Fritz Haber, the Nobel laureate who launched the age of chemical warfare* (New York: HarperCollins, 2005).

Much of the information on Fritz Houtermans recounted in Chapter 8 comes from my conversations with his son, Jan, which took place when we were both at the University of California, San Diego in the late 1960s. (Jan was later a professor at the University of Bern, where his father had held his last appointment.) The most complete published account of his remarkable life is in Thomas Powers' *Heisenberg's War,* pp. 84-112 (see also other entries in the book's index and Chapter 9, note 12, p. 502, for Powers' sources). The message that Houtermans sent in 1941 to the émigré scientists in the United States, saying that Heisenberg was trying to slow down any bomb project, will be found on pp. 106-107. The book that Houtermans and a colleague wrote using pseudonyms, and originally published in German, is: F. Beck and W. Godin, *Russian Purge and the Extraction of Confession,* translated by Eric Mosbacher and David Porter (New York: Viking Press, 1951). It has long been out of print, of course, but one can still find copies from second-hand booksellers through Amazon.

Those interested in "The Evolution of Placental Mammals" (Chapter 10) should consult above all else the great book by Sarah Blaffer Hrdy, *Mother Nature: A history of mothers, infants, and natural selection* (New York: Pantheon Books, 1999). Also useful is Louann Brizendine, *The Female Brain* (New York: Morgan Road Books, 2006). The epigraph for Part Three can be found in the fine article by Cort A. Pedersen, "How love evolved from sex and gave birth to intelligence and human nature," *Journal of Bioeconomics*, vol. 6 (2004), pp. 39-63; the sentence quoted is on p. 48. On *rattus*, see Natalie Angier, "Smart, Curious, Ticklish. Rats?" *The New York Times*, July 24, 2007, and A. L. Foote and J. D. Crystal, "Metacognition in the Rat," *Current Biology*, vol. 17 (2007), 551-555.

CHAPTER 12:

(1) Material on both the Franck Report and "Ötzi the Iceman" can easily be found on the Internet.

(2) The passage in the quotation that begins, "There are two ways of dealing with dangerous technologies," is attributed to MIT's Tom Knight and is cited in the May 20-26, 2005 issue of *New Scientist*, p. 46. The author wishes to apologize to the eminent Dr. Knight for the disrespectful comment made by Hera about this passage.

(3) The complete report entitled "Ignition of the Atmosphere with Nuclear Bombs" (1946), prepared by F. von Konopinski, C. Marvin, and E. Teller, was declassified (approved for public release) by Los Alamos National Laboratory in 1979, but was then officially withdrawn again following the incidents of September 11, 2001. However, one can still easily find the document in its entirety on the Internet.

(4) The Brookhaven experiment using the relativistic heavy ion collider is also known as the "killer strangelet" disaster scenario, and one can find information about it

on the Internet, using that phrase. For fascinating discussion of this risk scenario, see the article by Adrian Kent, "A critical look at risk assessments for global catastrophes" (2003), available on the Internet.

Chapter 15:

Of all the contentious issues related to the post-2001 "war on terror," none is more bizarre than the sinister terrorist attack on US soil involving weaponized anthrax. This attack, launched by material distributed in letters through the US postal system, occurred in October 2001, mere weeks after September 11, eventually caused 22 cases of inhalation anthrax, five of which were fatal. See D. B. Jernigen et al., "Investigation of bioterrorism-related anthrax, 2001: Epidemiologic findings," *Emerging Infectious Diseases,* vol. 8, no. 10 (October 2002). At the time of writing (December 2007), more than six years had elapsed since the events, and no one has ever been arrested for the crime; there is persistent speculation that someone who had worked at the US Army's bioweapons facility at Fort Dietrich, Maryland, must have been involved. The President of the United States never mentions this unsolved mystery in the updates on the war on terror. Information available on the Internet from reliable sources shows that the material used in these attacks had an exceptionally high spore count (1 trillion per gram); in addition, its method of preparation showed a level of sophistication that could not have been attained by any "amateur" group. See, for example: Gary Matsumoto, "Anthrax Powder: State of the Art?" (*Science,* vol. 302, no. 5650 [28 Nov. 2003, pp. 1492-1497]). There are superb illustrations in this article.

§ § §

The passage from the second section of the "Offertorio" in Giuseppe Verdi's *Requiem Mass* (composed in 1873), referred to toward the end of Chapter 9, is in the Latin:

Hostias et preces tibi, Domine
laudis offerimus,
tu suscipe pro animabus illis,
quarem hodie memoriam facimus:
fac eas, Domine, de morte
transire ad vitam,
quam olim Abrahae promisisti
et semini eius.

The recording featuring the tenor Jussi Björling is on London CD 444 833-2; the English translation in the text is borrowed from the notes that accompany this version. An LP containing selections from Björling's roles, including the "Ingemisco" from Verdi's requiem, was one of the first records I owned long ago.

Recently, however, my old friend David Ober, who had tutored me in classical music when we were both graduate students at Brandeis University in the early 1960s, persuaded me to pay more attention to the rendition conducted by Sir John Barbirolli (EMI Classics, packaged with Daniel Barenboim conducting Mozart's requiem), advice for which I am most grateful. The soloists on the Barbirolli recording include the Canadian tenor Jon Vickers, along with Fiorenza Cossotto (mezzo) and Ruggero Raimondi (bass); the performance by the soprano, Montserrat Caballé, is simply transcendent.

Gustav Mahler (1860-1911) began composing his song-cycle *Das Lied von der Erde* in 1907; "Der Abschied" ("Farewell") is the sixth and last of the songs and is written for mezzo-soprano. (There is a discussion of this work in *Hera, or Empathy* at pp. 593-596, which errs in stating that there is no CD available featuring the great Canadian mezzo Maureen Forrester; a live recording from Berlin in 1967, with George Szell conducting, can be found on "Living Stage" LS 1053.) The original German for the passage cited as the epigraph for Part Four (in my translation) is as follows:

Die Blumen blassen im Dämmerschein.
Die Erde atmet voll von Ruh' und Schlaf,
Alle Sehnsucht will nun träumen.
Die müden Menschen geh'n heimwärts,
Um im Schlaf vergeß'nes Glück
Und Jugend neu zu lernen!

Some background for Hera's discussion of this work in Chapter 2 may be found in Stuart Feder's *Gustav Mahler: A Life in Crisis* (New Haven: Yale University Press, 2004).

For more on Beethoven's medical conditions, mentioned in Chapter 14, see François Martin Mai, *Diagnosing Genius: The Life and Death of Beethoven* (Montreal: McGill-Queen's University Press, 2007).

ABOUT THE HERASAGA

BOOK ONE: *Hera, or Empathy*
BOOK TWO: *The Priesthood of Science*
BOOK THREE: *Hera the Buddha*

Thematic Outline:
The Way of Reflection on mind's relation to nature passes through the moments of submission (religion) and dominion (technology) toward its goal—mind's peace with nature.

Since the beginnings of human civilization 6,000 years ago in the Near East—in Egypt and Mesopotamia—Mind (human thinking) has been at war with Nature in two vastly different but complementary forms, namely, religion and technology. In both of these forms, Nature is nothing in itself, simply a background field of matter and energy onto which human meaning and power is projected and imposed.

Represented systematically, this process develops as follows:

Positing: The religious representation of reality posits Nature passively, not self-originating, as created by Absolute Spirit or God, in which Mind participates derivatively as Soul.

Negation: In order to fulfill itself as technology, Mind posits Nature as the Other to itself, merely "mindless" matter and energy governed by laws, and in this way finally unlocks the secrets of its own self-origins as Nature (that is, as the product of DNA's evolution).

Negation of negation: Mind dissolves Spirit and understands itself as natural, as a product of nature, and thus as limited in time, not infinite or absolute.

BOOK ONE. HERA, OR EMPATHY:

The cover artwork, Colville's "Church and Horse" (1964), shows nature in opposition to the religious representation of reality, in which nature is the passive outcome of an act of creation. Here religion is portrayed as a lifeless empty façade (the building) and broken domain (the gate), and is juxtaposed to the fierce energy and determination of the living, riderless animal that moves menacingly toward the standpoint of the viewer, unstoppably away and out.

BOOK TWO. THE PRIESTHOOD OF SCIENCE:

The cover artwork, Colville's "Horse and Train" (1954), shows human technology in its head-on confrontation with a vastly more powerful, living nature. The horse, again riderless and uncontrolled, moving away from the viewer, paradoxically towering in size over the train and opposing itself fearlessly to it, sets itself squarely upon the tracks, eschewing the surrounding fields.

BOOK THREE. HERA THE BUDDHA:

The cover artwork, Colville's "Moon and Cow" (1963), shows a much-domesticated animal—which is thus itself both a human creation yet still also living nature—at rest in the night, after a long day of her labors in the service of human needs (but with no human masters now present), facing away from the viewer, at peace with nature.